Studies in Big Data

Volume 15

Series editor

Janusz Kacprzyk, Polish Academy of Sciences, Warsaw, Poland
e-mail: kacprzyk@ibspan.waw.pl

About this Series

The series "Studies in Big Data" (SBD) publishes new developments and advances in the various areas of Big Data- quickly and with a high quality. The intent is to cover the theory, research, development, and applications of Big Data, as embedded in the fields of engineering, computer science, physics, economics and life sciences. The books of the series refer to the analysis and understanding of large, complex, and/or distributed data sets generated from recent digital sources coming from sensors or other physical instruments as well as simulations, crowd sourcing, social networks or other internet transactions, such as emails or video click streams and other. The series contains monographs, lecture notes and edited volumes in Big Data spanning the areas of computational intelligence incl. neural networks, evolutionary computation, soft computing, fuzzy systems, as well as artificial intelligence, data mining, modern statistics and Operations research, as well as self-organizing systems. Of particular value to both the contributors and the readership are the short publication timeframe and the world-wide distribution, which enable both wide and rapid dissemination of research output.

More information about this series at http://www.springer.com/series/11970

Ana Paula Ferreira Dias Barbosa Póvoa
João Luis de Miranda
Editors

Operations Research and Big Data

IO2015-XVII Congress of Portuguese
Association of Operational Research (APDIO)

Springer

Editors
Ana Paula Ferreira Dias Barbosa Póvoa
Instituto Superior Técnico
Lisboa
Portugal

João Luis de Miranda
Instituto Politécnico de Portalegre
Portalegre
Portugal

ISSN 2197-6503 ISSN 2197-6511 (electronic)
Studies in Big Data
ISBN 978-3-319-24152-4 ISBN 978-3-319-24154-8 (eBook)
DOI 10.1007/978-3-319-24154-8

Library of Congress Control Number: 2015949308

Springer Cham Heidelberg New York Dordrecht London

Springer International Publishing AG Switzerland is part of Springer Science+Business Media
(www.springer.com)

Preface

It is our great pleasure to present this volume of "Operations Research and Big Data", as the proceedings of the event IO 2015, the XVII Congress of the Portuguese Association of Operational Research, APDIO, held this year in Portalegre, Portugal. Through this conference we would like to highlight the contributions of the Portuguese Operational Research community to society with the focus of Operational Research and Big Data.

This conference serves as a forum for scientific community and practitioners of operational research to discuss the current challenges that need to be addressed, and, to highlight new developments in the applications of methods, algorithms, and tools to solve a wide range of both real world and theoretical problems.

IO 2015 includes three keynote lecturers from well-known experts in the area, Ayse Basar Bener (Ryerson University, Canada), António Pais Antunes (University of Coimbra, Portugal), and José Fernando Oliveira (University of Porto, Portugal). In addition close to 95 works will be presented, from which 29 papers have been accepted and published in this volume. These papers cover wide range of Operational Research topics and have been peer-reviewed by the scientific committee. We thank the scientific committee members for their timely and thorough reviews. We thank the IO 2015 Organization Committee for the materialization of this volume and finally, we also want to thank all the authors for their high quality manuscripts and for submitting them on time.

We do hope the contents of this volume will serve as valuable reference to the scientific community and practitioners of operational research and a permanent record of the IO 2015 conference.

Ana Paula Barbosa Póvoa
João Miranda
Susana Relvas
(co-editor)

Contents

A Model to Minimize Costs and Promote Species Persistence under Climate Change

Diogo Alagador[1] and Jorge Orestes Cerdeira[2]

[1] CIBIO/InBio-UE: Centro de Investigação em Biodiversidade e Recursos Genéticos,
Universidade de Évora, Évora, Portugal
alagador@uevora.pt

[2] Departamento de Matemática and Centro de Matemática e Aplicações,
Faculdade de Ciências e Tecnologia, Universidade Nova de Lisboa,
Costa da Caparica, Portugal
jo.cerdeira@fct.unl.pt

Abstract. Biodiversity is severely threatened by the effects of changing climates that cause species to readjust their spatial ranges. But species are limited in their capacity to follow suitable climates, as these rearrange in space along time. Therefore, the identification of the areas to support spatial readjustments of species is a pivotal step that should be made thoroughly given the limited budgets available. We propose a two-stage mixed integer linear programming to formalize this issue, present a heuristic and report results comparing optimal and heuristic solutions.

1 Introduction

Biodiversity has been threatened by multiple factors of distinct nature and magnitude. In order to make biodiversity to persist, conservation science urges to direct efforts to identify areas, known as reserves or protected areas (PA), capable to effectively safeguard biodiversity from their most impacting threats (Araújo et al. 2011).

In this chapter we briefly report models proposed to achieve efficiency and effectiveness in area selection under stressing changing climates, pointing out their main shortcomings. We present an alternative spatial conservation model that overcomes these issues, for which we give a mixed integer linear programming (MIP) formulation and present a heuristic. Results are given comparing optimal and heuristic solutions using several synthetic datasets.

2 Previous Works and Unresolved Issues

Several studies have been conducted to identify sets of areas likely to cover species in suitable climates, but in most of them species' abilities to follow the pace of climate change are not considered (or are accounted very simplistically).

A.P.F.D. Barbosa Póvoa and J.L. de Miranda (eds.), *Operations Research and Big Data,*
Studies in Big Data 15, DOI: 10.1007/978-3-319-24154-8_1

Williams et al. (2005) pioneered the integration of the capability of species to disperse within spatial prioritization models considering climate change, by means of the identification of sets of time-based sequence of areas to assist species' movements in their climatic adaptation. They call these conservation units, dispersal corridors and developed a heuristic to minimize the total area occupied by a given number of corridors for each species. Using data on species occurrences and corresponding climatic patterns under forecasted climate scenarios, they identified the areas where suitable climates for each species will occur in future. Williams et al. (2005) also accounted for data on species' dispersal abilities in order to predict what regions among the climate suitable ones a species may potentially colonize.

Later Phillips et al. (2008) formulated this same conceptual framework as a network flow problem. Their model considers, for each species, a network describing feasible dispersal corridors between consecutive periods of time. Figure 1 illustrates one such network for three time periods t=1,2,3. Nodes of the network in each period t indicate areas where the species is present in time t. For technical reasons, that will be clarified latter, each area is represented by a pair of nodes u, u' which are linked by arc (u,u'). An arc (u,v) between consecutive time-periods indicates that the species has capacity to move from area u to v within these time periods. An extra node (source) and arcs connecting this node with the first copy of the areas of time t=1 are added, and an extra node (terminal) and arcs connecting every second copy of the areas of the last period to this node are added too. A path from the source to the terminal nodes identifies a feasible corridor for species to disperse with time. The problem that Williams et al. (2005) and Phillips et al. (2008) considered consists of selecting a given number of node-independent paths (i.e., feasible corridors not intersecting in the same time-period) for each species so to minimize the total number of nodes on these paths. A feeble point of this model lies on the assumptions that i) a species is either present or absent in each of the areas, and ii) given two areas where a species is present in two consecutive time periods, either species can move between the two areas or not. As data on distribution of species often comes from models that assign to each area a value quantifying the suitability of that area for each species, a yes/no decision on the occurrence of species seems rather simplistic. Moreover, it is also doubtful to know exactly if a species can or not move between two arbitrary areas within a certain period of time. To overcome these limitations, taking advantage on data retrieved by niche models and information on species-specific dispersal, Alagador et al. (2014) defined a persistence (probabilistic) index for a species s to disperse along a corridor cor=$(v1,v2,..., vm)$, consisting of area vt in period t, i=1,..., m, as:

$$p_s(cor) = po_s^{v1} \times pd_s^{v1,v2} \times po_s^{v2} \times pd_s^{v2,v3} \times ... \times pd_s^{v(m-1),vm} \times po_s^{vm} \quad (1)$$

where po_s^{vt} is the occurrence probability of species s on area vt on time period t and $pd_s^{vt,v(t+1)}$ is the probability of a successful dispersal of species s from vt to

$v(t+1)$. Alagador et al. (2014) considered the problem of finding a predefined number of corridors for each species that maximizes the product of $p_s(cor)$ for all the corridors defined for all species, with a bound on the number of selected areas.

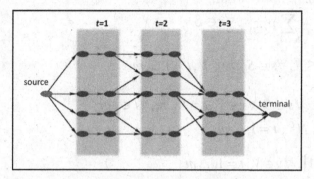

Fig. 1 A network defining the spatial distribution of a species along time in a region. Each pair of nodes connected by an arc, in each time period, represents an area where the species is predicted to occur. Arcs linking nodes between consecutive time periods represent the capacity of the species to disperse within the corresponding time interval

While this model appears more realistic (although at expenses of computational efficiency) than the one of Williams et al. (2005) and Phillips et al. (2008), it may produce solutions that neglect the persistence of some species. Another issue that both models bear, which seems little natural from a conservationist point of view, is the need to indicate a pre-established number of corridors for each species. It is certainly more significant to find estimates for adequate levels of species persistence (Araújo and Williams 2000; Williams and Araújo 2000; Justus et al. 2008).

In this paper we propose a strategy to resolve these issues. The strategy consists of two stages. The first, aims to maximize persistence values for those species whose desirable levels of persistence are not possible to achieve with area limitation. The second departs from these optimized persistence targets and identifies sets of corridors with minimum total cost under the established area limitation.

3   The Proposed Model

As defined above a corridor is an ordered sequence $cor=(v1, v2,\ldots,vm)$ of m (possibly not distinct) areas assigning to each time period $t=1,\ldots,m$, an area vt. The persistence of species s along corridor cor is given by expression (1). For each species s, let cor_s be the set of corridors with non-zero persistence, cor_s^{vt} be the subset of corridors of cor_s, which includes area v at time period t, and let P_s be the desirable persistence for species s, viewed as the sum of species s persistence along corridors in cor_s. We use S to denote the set all species, V the set of all areas, and N^t the maximum number of areas that can be selected in each time-period t.

The first stage of our model (MIP1) defines values for variables y_s in order to:

$$\min \sum_{s \in S} y_s \tag{2}$$

Subject to:

$$y_s \geq P_s - \sum_{l \in cor_s} p_s^l . z_s^l \; , \forall s \in S \tag{3}$$

$$\sum_{l \in cor_s^{v,t}} z_s^l \leq 1 \; , \forall s \in S, \forall v \in V, t = 1,...,\dot{m} \tag{4}$$

$$x_v^t \geq z_s^l \; , \forall s \in S, \forall v \in V, t = 1,...m, \forall l \in cor_s^{v,t} \tag{5}$$

$$\sum_{v \in V} x_v^t \leq N^t \; , t = 1,...,m \tag{6}$$

$$x_v^t \in [0,1], \; \forall v \in V, t = 1,...,m \tag{7}$$

$$z_s^l \in \{0,1\}, \; \forall s \in S, \forall l \in cor_s \tag{8}$$

$$y_s \geq 0, \; \forall s \in S \tag{9}$$

Variables z_s^l indicate whether corridor $l \in cor_s$ is selected ($z_s^l=1$) or not ($z_s^l=0$). Inequalities (3) define y_s to be at least the difference between the desirable persistence established for species s and the persistence obtained from the selected corridors. If $y_s=0$ it means that the corridors do satisfy the required persistence for species s. Inequalities (4) state that no more than one corridor from $cor_s^{v,t}$ is accounted in order to evaluate the persistence of species s. In other words, to compute the sum of persistence values for each species only corridors not intersecting in the same time-period (i.e., node-independent) are considered. Inequalities (5) assign to variables x_v^t value 1 if some corridor l is selected having area v in period t, i.e., $z_s^l=1$. Inequalities (6) guarantee that, on every time period t, the sum of variables x_v^t does not exceed the maximum number of areas to be selected for that time-period. Constraints (7-9) define the range of variables. The objective function (2) seeks to minimize the sum of gaps of persistence y_s. Given the way variables y_s, x_v^t, and z_s^l are related, in every optimal solution variables x_v^t take only zero-one values, with $x_v^t=1$ indicating that area v is selected during time period t.

In the second stage (MIP2) the model uses the values $\hat{y}_s$ assigned to variables y_s in an optimal solution of MIP1. Given the costs c_v^t of using for conservation purposes area v during the time period t, MIP2 consists of

$$\min \sum_{t=1,...,m} \sum_{v \in V} c_v^t . x_v^t \tag{10}$$

Subject to:

$$\sum_{l \in cor_s} p_s^l . z_s^l \geq P_s - \hat{y}_s, \; \forall s \in S \tag{11}$$

and constraints (4-9).

Inequalities (11) ensure that the persistence of each species will be at least the feasible persistence value determined from MIP1 solution, imposing the number of areas in each period t to be no more than N^t. The objective function (10) searches a solution of minimum cost.

Both MIP1 and MIP2 use variables z_s^l associated to every corridor, $l \in cor_s$, the set of all corridors for which species s has positive persistence, whose size may be huge making the formulations computationally impractical. To handle this we replace set cor_s by its subset $côr_s$ of the k_s corridors with the greatest persistence for species s. Although NP-hard, defining $côr_s$ can be quickly achieved for moderately large values of k_s, using for instance the algorithm in Martins et al. (1999).

4   A Heuristic

Given that real conservation problems often apply to very extensive and resolute datasets, MIP solutions will likely be computationally hard to obtain, even if replacing cor_s by much smaller subsets $côr_s \subseteq cor_s$. We therefore developed a simple heuristic to handle such limitation.

The heuristic works as follows. Starting with the empty set C, in each step the heuristic randomly selects a species s for which the persistence target P_s is not achieved by the corridors in C, and adds to C the highest-ranked corridor, cor, not yet selected in $côr_s$. To check whether C achieves P_s it should be taken into account that to evaluate the persistence of a set of corridors for a species only corridors not intersecting in the same time-period (i.e., node-independent corridors) should be considered. To deal with this the algorithm maintains a graph G, where nodes correspond to corridors of C and where the existence of an edge $\{l,n\}$ indicates that corridors l and n are not node-independent. Clearly, node-independent corridors of C correspond to stable sets of G. Now, for each species s assign to every node l of G a weight equal to the persistence of s in the corridor corresponding to l. In this way we obtain for every species s a node-weighted graph G_s. To compute the persistence of the set of corridors C for species s we search for the maximum weighted stable set of G_s, which we do using a greedy randomized adaptive search procedure (Feo and Resende 1995).

The heuristic repeatedly proceeds adding corridors cor to C as above described, until the desirable persistence for all species are satisfied or adding cor to C the maximum area allowed for some time period, N^t, is exceeded. When this happens phase one of the heuristic is completed. Note that when a corridor is included in C, graph G is updated and maximum weighted stable set of graph G_s is computed for every species s for which the required persistence target P_s is not achieved by C. At the end, each species s has a persistence associated to C, denoted by P_s^C, which is either greater or equal to P_s, or equal to the weight of a maximum-weighted stable set of graph G_s. The heuristic now proceeds removing corridors from C accounting for the costs of their areas while maintaining P_s^C, for every species s. In each step of this pruning phase, a time period t is randomly selected and, among the areas included in the corridors of C in period t, an area v is selected with probability proportional to its cost. Thus more costly areas are more likely to be selected. Next, a procedure will check whether removing from C all corridors that include v in time period t will decrease P_s^C. If this happens for at least one species, the corridors are maintained in C, and area v is marked as mandatory in time-period t. Otherwise, the corridors are removed from C, and graph G is updated. Note that to decide if the corridors are to be removed requires solving a maximum weighted stable set problem on every graph G_s. The whole process is repeated until all areas among the m time period are addressed.

The heuristic was intentionally designed to be of random nature as this allows identifying different solutions from different runs, delivering to decision makers a pool of alternative solutions to examine and compare.

5   Computational Results

Several distinct synthetic datasets were built varying some parameters: number of species, $|S|$=10 and $|S|$=50; map size, V=10 x 10, V=25 x 20 and V=25 x 40 grid cells; number of pre-selected corridors per species, k_s=50 and k_s=200 corridors; maximum conservation area allowed per time-period, N^t=.1V, and N^t=.5V of mapped grid cells; and initial required species' persistence targets, P_s=X and P_s=X/10, where X was obtained from a $N(1;1)$ Gaussian distribution.

We simulated suitability maps for m=4 time-periods using a random field function to define suitability scores for each species among grid cells with time. The maximum number of consecutive adjacent cells that the species is able to move between consecutive time-periods (i.e. dispersal rate), $disp_s$, was an integer uniformly selected in interval [1,15]. The probability of successful dispersal was established using a linearly decaying function with $pd_s^{vt,v(t+1)}$=1 if distance from vt to $v(t+1)$ is zero, and $pd_s^{vt,v(t+1)}$=0 if distance from vt to $v(t+1)$ is $disp_s$ or higher.

We assigned unit-costs to every grid cell for all time periods. Thus the cost of a solution is the sum of the number of cells in each time-period that are required to delineate dispersal corridors.

We ran the heuristic and evaluated the performance of the feeding and the pruning phase solutions, taken independently. We compared persistence gaps P_s-y_s among the optimal solutions of MIP1 with persistence gaps P_s- P_s^C obtained with the heuristic feeding-phase. Additionally we reported on the number of species for which the pre-established persistence targets were attained (covered species) because it is also an ecologically-important informer (Fig. 2a). We also compared the costs of optimal solutions of MIP2 with those obtained from the pruning phase of the heuristic, both fitted with the adapted persistence targets from the heuristic's feeding-phase (Fig. 2b). Finally, we contrasted the heuristic solutions with the ones retrieved by MIP2 using $\hat{y}_s$ from MIP1, using two criteria: the optimized species persistence targets, and the solution costs.

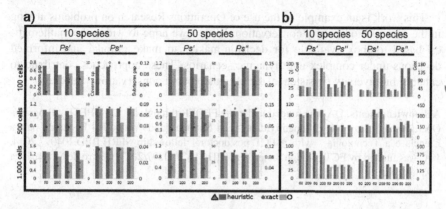

Fig. 2 Comparative performance of heuristic and exact solutions in terms of a) ecological effectiveness: measured as summed persistence gaps of underrepresented species (bars) and number of covered species (triangles and circles), and b) solution cost (bars), among several simulations.

The heuristic and MIP formulations were programmed using C-programming, with calls to *Cplex 12.5.1 Solver* for solving MIP1 and MIP2.

The performance of the heuristic's second-phase was slightly better than the one from the first-phase, with zero to 88% (mean=22%) higher costs and 0.2% to 98% (22%) higher persistence gaps than MIP2 and MIP1 solutions, respectively.

When assessing the problems using MIP1 and MIP2 (fed with $\hat{y}_s$ from MIP1) distinct responses occurred when comparing with heuristic solutions: i) optimal solutions with more restrictive persistence targets (lower persistence gaps) and, accordingly, higher solution costs than heuristic solutions and, ii) even with higher settled persistence targets, optimal solutions were less costly than heuristic ones.

In terms of computational times, the k min-cost paths algorithm (Martins et al. 1999) ran in .016 sec when defining the k_s=200 corridors for the species with higher dispersal ability in the 25 x 40 grid cells map.

The longest heuristic run took 947 sec ($|S|$=50, V=25 x 40, N'=.5V, k_s=200) and compares with 1.18 sec obtained with MIP1. Contrarily, MIP1 solved the instance differing in N'=.1V from the previous one in 5,238.3 sec comparing with the 15.3 s obtained by the heuristic. Times were obtained using a *Microsoft Windows* 64-bit workstation with 28 Gb RAM and an *Intel* 2.40 Ghz processor.

6   Conclusion

We proposed exact and approximate approaches to identify priority areas for species accounting for climate change, both departing from a pool of top persistence ranked corridors for each species using a k-rank min-cost paths algorithm. We compared the heuristic with the corridor-constrained exact approach, with results showing low sub-performance of the former when compared to the latter.

This work is an example of the use of Operations Research on problems arising in the area of biodiversity conservation. With it we hope to widen the mathematical-based toolbox available for decision-makers to make rational and informed decisions under complex and dynamic scenarios. These approaches contribute to enhance biodiversity persistence, the ultimate goal of conservation science.

Acknowledgments. DA was funded under the FEDER through the COMPETE - Programa Operacional Factores de Competitividade - and National funds via FCT – Fundação para a Ciência e a Tecnologia - with a FCT postdoctoral fellowship SFRH.BPD.104077.2014. JOC was funded by FCT through the project UID/MAT/00297/2013, Centro de Matemática e Aplicações.

References

Alagador, D., Cerdeira, J.O., Araújo, M.B.: Shifting protected areas: scheduling spatial priorities under climate change. Journal of Applied Ecology 51(3), 703–713 (2014)

Araújo, M.B., Alagador, D., Cabeza, M., Nogués-Bravo, D., Thuiller, W.: Climate change threatens European conservation areas. Ecology Letters 14, 484–492 (2011)

Araújo, M.B., Williams, P.H.: Selecting areas for species persistence using occurrence data. Biological Conservation 96(3), 331–345 (2000)

Feo, T., Resende, M.C.: Greedy Randomized Adaptive Search Procedures. J. Glob. Optim. 6(2), 109–133 (1995)

Justus, J., Fuller, T., Sarkar, S.: Influence of representation targets on the total area of conservation-area networks. Conservation Biology 22, 673–682 (2008)

Martins, E.Q.V., Pascoal, M.M., Santos, J.L.: Deviation algorithms for ranking shortest paths. International Journal of Foundations of Computer Science 10(03), 247–261 (1999)

Phillips, S., Williams, P., Midgley, G., Aaron, A.: Optimizing dispersal corridors for the Cape Proteaceae using network flow. Ecological Applications 18, 1200–1211 (2008)

Williams, P., Hannah, L., Andelman, S., Midgley, G., Araujo, M., Hughes, G., Manne, L., Martinez-Meyer, E., Pearson, R.: Planning for climate change: Identifying minimum-dispersal corridors for the Cape Proteaceae. Conservation Biology 19(4), 1063–1074 (2005)

Williams, P.H., Araujo, M.B.: Using probability of persistence to identify important areas for biodiversity conservation. Proceedings of the Royal Society of London Series B-Biological Sciences 267(1456), 1959–1966 (2000)

A Fix-and-Relax Algorithm for Solving Parallel and Sequential Versions of a Multi-period Multi-product Closed Loop Supply Chain Design and Operation Planning Model

Susana Baptista[1], Maria Isabel Gomes[1], Laureano Escudero[2],
Pedro Medeiros[3], and Filipe Cabrita[3]

[1] Centro de Matemática e Aplicações, FCT, Universidade Nova de Lisboa,
Caparica, Portugal
[2] Estadistica e Investigacion Operativa, Universidad Rey Juan Carlos, Madrid, Spain
[3] Departamento de Informatica, FCT, Universidade Nova de Lisboa, Caparica, Portugal

Abstract. In this work we present the sequential and parallel versions of a heuristic algorithm for the solution of a two-stage stochastic mixed 0-1 model for closed loop supply chain planning problem along a time horizon. Some computational experience conducted on randomly generated networks shows the quality of the proposed approach.

Keywords: Closed Loop Supply Chain, Design and Operation Planning, Two-stage Stochastic Mixed 0-1 Optimization.

1 Introduction

The fact that simultaneous design of the forward and reverse channels may lead to significant cost savings, has focused the interest of industry and academia in closed loop supply chains network design. Still, the research in the field has mostly addressed the deterministic case. However, the network parameters are uncertain by nature along a time horizon. Some of the first works addressing the stochastic case are presented in [1] for a single product in a multi-period and in [2] for a single product network in single period. [3] presents a two-stage stochastic model for a multi-product network in single period. The multi-period setting has been recently addressed in [4] and [5]. A recent review on reverse logistics and closed loop supply chain presented in [6] provides a very deep perspective on the work developed in this field. In most published works, a limited number of sources of uncertainty is addressed (mostly, product demand) and the solving procedures are either based on exact commercial solvers that cannot tackle real size problems or on meta-heuristic procedures that are too problem dependent.

In this work we address several sources of uncertainty simultaneously and assume the related random vector to have a discrete distribution, so that the design

© Springer International Publishing Switzerland 2015

A.P.F.D. Barbosa Póvoa and J.L. de Miranda (eds.), *Operations Research and Big Data*,
Studies in Big Data 15, DOI: 10.1007/978-3-319-24154-8_2

and planning of a closed loop network is modelled as a two-stage stochastic mixed 0-1 program. Due to the large scale of the problem, a stochastic version of the heuristic Fix and Relax algorithm presented in [7] is introduced in this work to tackle the problem under study. The sequential and parallel versions of the algorithm are also presented. A computational experience is reported in a set of randomly generated networks, where both versions of the algorithm are considered.

2   Problem Description and Modelling Approach

The modelling framework subject of this paper is an extension of the deterministic closed loop supply chain model introduced in [8] whose risk management is presented in [9]. The forward and reverse supply chains are operated by the same original equipment manufacturer (OEM). Products recovered from clients (end-of-life products) and processed in disassembly centres, are sorted according to their quality: top quality products are sold to a secondary market, good enough products are sent back to plants to be remanufactured and low quality products are disposed. The overall network decisions concern strategic (network topology) and operational issues (network usage planning).

Several sources of uncertainty are taken into account, namely products' demand, returned products' qualities and volumes, transportation costs, available budget for attending the financial costs, annualized amortization of the investment costs in all entities, etc., where entities name plants, distribution and disassembly centres. Since the random vector ξ that pieces together the stochastic components was assumed to have finite support, all its possible realizations are fully described by a set of scenarios, say Ω.

We point out that due to the multi-period (e.g. 15 years) character of the problem and the different nature of the decisions involved, strategic decisions (i.e. location of the three-echelon network and the production capacities) must be specified at pre-defined periods (so named macro-periods) in the time horizon (e.g., years 1, 6 and 11), while all operational decisions (i.e. production and inventory levels and network flows) are to be taken for every period t of the time horizon (e.g., a year) given by the set T.

Let the following notation be considered for the two-stage model presented below: a_1 and b_1 denote the objective function coefficients for the 0-1 strategic variables in vector x_1, and the continuous (operation) variables in vector y_1, respectively, a_t^ω and b_t^ω are the related coefficient vectors for the x_t^ω and y_t^ω variables in period t, for scenario $\omega \in \Omega$, w^ω is the weight of scenario ω, A_1 and B_1 are the matrices for the first stage constraint, h_1 is the related rhs, and $A_1^{t\omega}$, $B_1^{t\omega}$, A_t^ω, B_t^ω, h_t^ω are the related elements for the second stage constraint. Then the two-stage model can be expressed by

$$\max \quad a_1 x_1 + b_1 y_1 + \sum_{\omega \in \Omega} w^\omega \sum_{t \in T} a_t^\omega x_t^\omega + b_t^\omega y_t^\omega$$

$$\text{s.t.} \quad A_1 x_1 + B_1 y_1 = h_1$$

$$A_1^{t\omega} x_1 + B_1^{t\omega} y_1 + A_t^\omega x_t^\omega + B_t^\omega y_t^\omega = h_t^\omega \qquad \forall t \in T, \omega \in \Omega \qquad (1)$$

$$x_1 \in \{0,1\}^{nx(1)}, y_1 \in \mathbb{R}^{ny(1)}$$

$$x_t^\omega \in \{0,1\}^{nx(t)}, y_t^\omega \in \mathbb{R}^{ny(t)} \; \forall t \in T, \omega \in \Omega$$

where the objective function is the maximization of the Net Present Value of the expected profit over the scenarios along the time horizon. Notice that the x variables represent the strategic decisions of location of all entities that compose the supply chain and the strategic decisions of plants' production capacity that are operated in the network. Since those decisions are to be taken from a discrete set of alternatives, the x variables are binary. The y variables represent the operational decisions involved, namely the volume of product flows between entities of consecutive levels (from plants to distribution centres, distribution centres to customers, customers to disassemble centres and from this last entity to secondary market, disposal and plants) and the inventory levels at all entities. Finally, the two sets of constraints represent, respectively, the several constraints that have to be ensured in time period 1 and for all periods and scenarios, namely balance equations among entities, upper and conditional lower bound on product flow between entities and on production at plants, stock upper bounding and annualized amortization bounding.

3 Algorithmic Approach

In the algorithm that we propose for solving model (1), a set of levels (denoted as L) is considered along the time horizon defined by disjoint sets of consecutive periods. At each level a mixed 0-1 model (1) is solved by fixing the binary variables of ancestor levels to the value obtained at the optimization of their related models and relaxing the integrality of the binary variables related to successor levels.

Let partition of level set L be such that $L = L_1 \cup L_2$ and $L_1 \cap L_2 = \emptyset$, where L_1 is the set of levels related to first stage and L_2 includes the other levels. We present a rough description of the algorithm where a reductive full backward step is imposed when the full time horizon has been covered. Such backward step reduces the number of entities that exhibit differentiated costs (in the case, plants) and, as pointed in [9], it leads to better results.

Let t^l denote the last period in level l, for $l \in L$ and M_l^ω the model for scenario $\omega \in \Omega$ at level l where $\omega=0$ if $l \in L_1$. Then, for any model M_l^ω the binary variables of ancestor levels are fixed (to the value, say $\hat{x}$) and the integrality of all binary variables of successor levels are relaxed.

Let E denote the set of entities and x_l^e denote the first stage binary variables related to entity e, for $e \in E$, where level l is set to 1. Then, the rough algorithm in its parallel and sequential versions is as follows:

Step 0: Set label *backward* := '*false*' and $z := -\infty$.
Step 1: (Solution for first stage levels).
Solve sequentially model M_l^0 $\forall l \in L_1$.
Set parameter $num\_e := \sum_{e \in E} \hat{x}_l^e$ (number of opened e entities).
Step 2: (Solution for second stage levels).
Parallel version:
Solve in parallel the $|\Omega|$ independent scenario-based models M_l^ω $\forall l \in L_2$
Sequential version:
Solve sequentially the $|\Omega|$ independent scenario-based models M_l^ω $\forall l \in L_2$.
Step 3: (Objective function value).
Compute z such that

$$z = \max (z, a_1\hat{x}_1 + b_1y_1 + \sum_{\omega \in \Omega} w^\omega \sum_{t \in T} a_t^\omega \hat{x}_t^\omega + b_t^\omega y_t^\omega).$$

If *backward* := '*false*' then set *backward* := '*true*', append constraint $\sum_{e \in E} x_1^e = num\_e\text{-}1$ to model M_1^0 and go to step 1.

4 Computational Experience and Results Analysis

The computational results for a set of multi-period multi-commodity networks randomly generated instances are reported as extensions of the deterministic case presented in [8] to the stochastic one. Four networks (N1,N2, N3, and N4) were generated considering 18 customers, Ω=12 scenarios and T=15 periods time horizon. Table 1 gives the models' dimensions. Its headings are: number of constraints (m), 0-1 variables ($n01$), continuous variables (nc), non-zero elements in constraint matrix (nel) and constraint matrix density (den, in %). Observe the large model dimensions of the instances considered.

Table 1 Models' dimensions

	m	$n01$	nc	nel	den
N1	200 578	159	91 100	783 797	0.0043
N2	369 190	309	173 735	1 516 202	0.0024
N3	706 414	609	339 005	2 975 097	0.0012
N4	823 036	618	386 663	3 360043	0.0011

For each network two instances were created by using two different sets of scenario's probabilities (namely, P1 and P2) as given in Table 2. Additionally, $|L|$=4, where L_l ={1,2,3} with $t^1 = t^2 = t^3 = 1$ and L_2 ={4} where t^4=15. Levels l=1, 2 and 3 share the first stage but differ in the entities to each they refer to, since priorities were set among them (by order: plants, distribution centres, and disassembly centres). The macro-periods are 1, 6 and 11. Note: it is assumed that the decisions on the plants selection can only be made at period 1.

Table 2 Scenario probabilities

	Sc1	Sc2	Sc3	Sc4	Sc5	Sc6	Sc7	Sc8	Sc9	Sc10	Sc11	Sc12
P1	0.01	0.01	0.03	0.10	0.10	0.10	0.15	0.25	0.25	0.15	0.05	0.04
P2	0.01	0.01	0.01	0.02	0.05	0.10	0.05	0.15	0.10	0.18	0.12	0.20

HW/SW platform: a WS with a 2 Intel Xeon E5430 266 GHz processor (4 cores each), 24 MB of RAM, *gcc* 4.9.2 as compiler, C++ code and CPLEX v12.6 as the MIP engine.

Table 3 shows the computational comparison of CPLEX and the sequential and parallel versions of the proposed algorithm, where the latter use 8 cores. The headings are: z_{CPLEX} solution value by plain using of CPLEX; z_{alg} proposed algorithm solution value, optimality *GAP* of z_{alg} versus z_{CPLEX}, such that, $GAP\% = (z_{CPLEX} - z_{alg}) / z_{CPLEX} \cdot 100$; t_{CPLEX}, t_{seqalg}, and t_{paralg} computing times (s.) for obtaining z_{CPLEX} and z_{alg} with the sequential and parallel versions of the algorithm, respectively.

Table 3 Computational results

	z_{CPLEX}	z_{alg}	$GAP\%$	t_{CPLEX}	t_{seqalg}	t_{paralg}
N1P1	25 739.9	25 739.9	0	25 653	6868	**2565**
N1P2	15 307.5	15 307.5	0	11 156	7788	**2778**
N2P1	210 684.5	210 684.5	0	**400**	2568	1085
N2P2	181 182.7	186 182.7	0	**170**	5301	3036
N3P1	-95 742.0	-95 742.0	0	768	1044	**272**
N3P2	-110 352.0	-110 352.0	0	457	1131	**342**
N4P1	221 680.7	221 680.7	0	**207**	1921	419
N4P2	151 899.0	151 899.0	0	**175**	1369	354

Observe that in all instances the algorithm produced the optimal solutions. It proved to be extremely effective in half of the instances, reducing the computing time up to one order of magnitude. On the other hand, CPLEX requires smaller computing times for those instances where the optimum is found at the very first B&B nodes, which is not very frequent in real world instances. Two major facts should be stressed. First, in all networks the parallel version of the algorithm achieved an impressive computing time reduction with respect to the sequential one, varying from 43 up to 78%. Second, the MIP problems solved at the first stage (that, by construction, is not parallelizable) are the hardest ones.

5   Conclusions

We have studied the performance of the sequential and parallel versions of a useful heuristic algorithm for the solution of a two-stage stochastic mixed 0-1 model of a closed loop supply chains design and operation planning problem in a dynamic setting where uncertainties appear anywhere.

Acknowledgement. This work was partially supported by the Portuguese National Science Foundation under the project UID/MAT/00297/2013.

References

[1] Inderfurth, K.: Impact of uncertainties on recovery behavior in a remanufacturing environment: A numerical analysis. Int. J. Phys. Distrib. 35, 318–336 (2005)

[2] Listes, O.: A generic stochastic model for supply-and-return network design. Comput. Oper. Res. 34(2), 417–442 (2007)

[3] Salema, M.I.G., Barbosa-Povoa, A.P., Novais, A.Q.: An optimization model for the design of a capacitated multi-product reverse logistics network with uncertainty. Eur. J. Oper. Res. 179(3), 1063–1077 (2007)

[4] Cardoso, S.R., Barbosa-Póvoa, A.P.F.D., Relvas, S.: Design and Planning of Supply Chains with Integration of Reverse Logistics Activities under Demand Uncertainty. Eur. J. Oper. Res. 226, 436–451 (2013)

[5] Zeballos, L.J., Méndez, C.A., Barbosa-Povoa, A.P., Novais, A.Q.: Multi-period design and planning of closed-loop supply chains with uncertain supply and demand. Comput. Chem. Eng. 66, 151–164 (2014)

[6] Govindan, K., Soleimani, H., Kannan, D.: Reverse logistics and closed-loop supply chain: A comprehensive review to explore the future. Eur. J. Oper. Res. 240(3), 603–626 (2015)

[7] Escudero, L.F., Salmeron, J.: On a fix-and-relax framework for a class of project scheduling problems. Ann. Oper. Res. 140, 163–188 (2005)

[8] Salema, M.I.G., Barbosa-Povoa, A.P., Novais, A.Q.: Simultaneous design and planning of supply chains with reverse flows: A generic modelling framework. Eur. J. Oper. Res. 203(2), 336–349 (2010)

[9] Baptista, S., Barbosa-Povoa, A.P., Escudero, L.F., Gomes, M.I.: A metaheuristic for solving large-scale two-stage stochastic mixed. programs with the time stochastic dominance risk averse strategy. In: 12th PSE and 25th ESCAPE (to appear, 2015)

Efficiency in Increasing Returns of Scale Frontier

Juliana Benicio[1], João Carlos Soares de Mello[1], and Lidia Angulo Meza[2]

[1] Universidade Federal Fluminense (UFF), Niterói, RJ, Brazil
juliana.benicio@hotmail.com, jcsmello@producao.uff.br
[2] Universidade Federal Fluminense (UFF); Volta Redonda, RJ, Brazil
lidia@metal.eeimvr.uff.br

Abstract. The main objective of this paper is to analyze DMUs efficiency from different perspectives of variable returns to scale. For that, the authors use results of CCR, BCC and a new model proposed. The new model presents just in-creasing returns to scale. Thus, the efficient frontier must have specifics characteristics that guarantees increasing additional output beyond additional inputs verified. This study is going to show that the classic model of DEA BCC proposes variable returns of scale, however cannot ensure that all the efficient DMUs with increasing returns to scale is identified. The authors propose a new model where the measurement of efficiency is made by increasing returns to scale frontier.

Keywords: Variable scale, increasing returns to scale, new DEA model.

1 Introduction

The classical definition of efficiency in economic science says that a production unit is fully efficient, if and only if, you cannot improve any output without im-prove any other input or reduce any input without reduce any output [13-14]. Thus, one unit is considered inefficient, if it can produce the same output reducing at least one of the inputs; or if you can use the same inputs to produce more out-puts. In this sense, the construction of a production line, also referred to as effi-cient frontier, aims to define a limit where the production more efficient will be located on this boundary, and the less efficient will be situated in the below border area, known as a set of possible production [21].

The shape of the efficient frontier defines the technology used in production analyzed. [8] points out that the efficiency measure must incorporate a set of pro-duction possibilities to maximize the ratio of output/input.

In the original paper [7], Charnes, Cooper and Rhodes used efficiency concept in linear programming model known as DEA CCR. The DEA (Data Envelopment Analysis) is able to evaluate the level of efficiency of production units (DMUs - Decision Making Units) that perform the same activity, in the use of its resources.

© Springer International Publishing Switzerland 2015

A.P.F.D. Barbosa Póvoa and J.L. de Miranda (eds.), *Operations Research and Big Data,*
Studies in Big Data 15, DOI: 10.1007/978-3-319-24154-8_3

The measure of efficiency is obtained by the reason of weighted sum of outputs by the weighted sum of inputs. This model allows to analyze performance DMU to produce multiple outputs from multiple inputs through [9] compared to the other DMU observed.

The result of the original model CCR is to build an efficient production frontier. The DMUs that have the best ratio "product/input" are considered more efficient and will be located on this border, and the less efficient will be situated in below the border region, known as envelope [15].

As analytical progress, efficiency studies began to incorporate the concept of scale in their analysis. That is, by increasing the volume of production inputs, DMUs considered efficient may have increase, decrease or constancy of productivity.

[2] BCC DEA developed a model that incorporates the notion returns to scale [3] to the efficiency analysis. As argued by [8] is a mistake to consider only the best productivity measure (better relationship between inputs and outputs) as a measure of efficiency. Thus, in the BCC model, the condition of proportionality of the CCR is not guaranteed in this model. The efficient frontier begins to recognize the possibility of improvement or deterioration of the DMU productivity by altering the amount produced [20].

This article aims to analyze different models in the definition of efficient DMUs from the concept of variable returns to scale. These different models presents different formats of frontiers. The proposed new model has a concave up frontier and the efficiency is measured by increasing returns to scale.

Looking for that, the article is going to present the theoretical foundation of the models: CCR, BCC and FConc. A comparison of each format of the efficient frontiers are going to be made. In the conclusion of the analysis will be presented the strengths and weakness of each model.

2   A Revision of Variable Returns of Scale: An Economic Perspective

The average productivity of a factor (PM) is calculated as the quotient of the output $f(x)$ and the amount used of the input in question (x). Algebraically:

$$PM = \frac{f(x)}{x} \tag{1}$$

To analyze the return to scale return a small increment of x is applied, and the variation of PM is verified (Miyazaki, 2001). So if:

- Increasing return to scale (IRS): $\frac{\Delta PM}{\Delta x} > 0$

- Decreasing returns to scale (DRS): $\frac{\Delta PM}{\Delta x} < 0$ (2)

- Constant returns to scale (CRS): $\dfrac{\Delta PM}{\Delta x} = 0$

Figure 1 shows the scale returns from possible change in the amount of inputs used, where the different behaviors of the efficient frontier curve production can be verified [18].

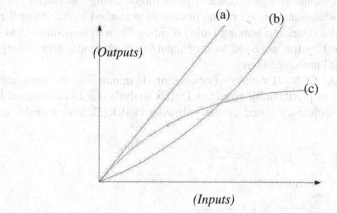

Fig. 1 (a) Constant returns of scale; (b) Increasing returns to scale; (c) Decreasing returns to scale

[3] highlight the following reasons that contribute to the technology shows increasing returns to scale:

- Existence of indivisibilities techniques or fixed costs, which are diluted with increasing production scale (eg costs of mobile telephone network, product de-sign, music or films).
- The division of labor and specialization can enable efficiency gains (eg production line).
- Inventory needs usually increase less than the scale (eg hypermarkets).
- Geometric relationships: for example, duplicating the walls of a warehouse, quadruples the available area.

[3] described reasons that contribute to the existence of decreasing returns to scale:

- Excess of work division and loss of overview of the company and its objec-tives (fruit of great organizational complexity);
- Supervisory difficulties/management: As the scale of production increases, the supervisors hierarchy tends to increase their efficiency and decreasing (also the result of major organizational complexity); Product limitation (extractive industries).

3   Studied Models

3.1   DEA CCR Model

The DEA CCR model introduced by [7] assumes constant returns to scale, meaning any change in inputs should produce a proportional change in output. The model uses the mathematical programming optimization method to determine the efficiency of a DMU (Decision Making Units) dividing the weighted sum of outputs (virtual output) by the weighted sum of inputs (virtual input), generalizing thus definition of [8] presented above.

The classic model CCR [7] with input orientation (ie minimizing the input and maintains the level of production), considers DMUs analysis unit to be compared according to their efficiency based on the following model (3), called model of Multipliers:

$$Max \; Eff_0 = \frac{\sum_{j=1}^{s} u_j y_{jo}}{\sum_{i=1}^{r} v_i x_{io}}$$

Subject to

$$\frac{\sum_{j=1}^{s} u_j y_{jk}}{\sum_{i=1}^{r} v_i x_{ik}} \leq 1, k = 1, \ldots, n \tag{3}$$

$$u_j \, e \, v_i \geq 0 \; \forall \, j, i$$

Such that, uj e vi are the weights of outputs and inputs, respectively; xik, yjk are the inputs ie outputs j of DMUK and xi0, yj0 are the inputs I and outputs j of the observed DMU.

This model can be defined as a fractional programming problem that can be transformed into a linear programming (LPP), where the denominator of the objective function must necessarily be equal to a constant, usually one [5]. The model (3) can be instructed to output, and thus, its objective function seek to maximize output, holding constant the level of inputs [6].

3.2   DEA BCC Model

The BCC model, introduced by Banker, Charnes and Cooper [3] introduced a change in the formulation of CCR in order to analyze the variable returns to scale in DEA. That is, the BCC model wanted to give account to interpret the fact that, at different scales, the DMUs could have different productivities and still be considered efficient. The objective of this analysis proposed by the BCC model is to take into account the fact that in different situations the conditions that influence the productivity of production are also diverse. As mentioned in the introduction, different production technologies have their productivities influenced by the scale at which the DMUs are operating [2]. When the production frontier exhibits constant returns to scale, efficient DMUs have the same productivity; however, when the production line has variable returns efficient DMU need not have the same productivity [12].

The formulation (4) of the BCC introduces a restriction on the PSS of the original model CCR. The frontier of this concave down set is restricted by $\sum_{i-1}^{n} \lambda_i = 1$, making the area BCC production possibilities less that the CCR. Consequently, any projection inefficient DMU in the efficient hyperplane, may be represented by an equation of the line segment of the linear combination border, **where the sum of the contributions of efficient DMUs (λ_j) must result in 1** [3]. Thus, the BCC efficiency is less than or equal to the CCR efficiency.

The model also introduces the variables v * (scale factor in the output orientation) and u * (scale factor in the input orientation) to the objective function and constraints. These variables, according to [2] indicate the scale return of the DMU. In the oriented to inputs model, when the scale factor is positive, an increasing return to scale is verified; when negative, an decreasing return is verified; and, finally, when if the scale factor is null, a constant return is verified. Followig is the model (4) of DEA BCC, input orientation:

$$Max \; Eff_0 = \frac{\sum_{j=1}^{s} u_j y_{jo}}{\sum_{i=1}^{r} v_i x_{io}} - u^*$$

Subject to

$$\frac{\sum_{j=1}^{s} u_j y_{jk}}{\sum_{i=1}^{r} v_i x_{ik}} - u^* \le 1, k = 1, ..., n \tag{4}$$

$$u_j e \; v_i \ge 0 \; \forall \; j, i$$

Such that, uj e vi are the weights of outputs and inputs, respectively; xik, yjk are the inputs i and outputs j of DMUk and xi0, yj0 are the inputs i and outputs j of the observed DMU; u* is the scale factor.

3.3   New Model: Concave Up Frontier (FCON)

This nonparametric algorithm is designed to ensure that the efficient frontier shows increasing returns to scale. The shape of the efficient frontier in this model is concave up. The concave up format ensures that the productivity increases are verified along the frontier.

To guarantee the increasing returns to scale, the efficient DMUs must present increasing CCR efficiency. Alternatively, the productivity of the efficient DMUs most increase constantly. Consider the following conventions:

• The DMUs are ordered by the numbers of inputs;
• Eff$_{CCR \; O}$ is the CCR Efficiency of the observed DMU;
• Eff$_{CCR \; O}$ is the CCR Efficiency of ALL efficient DMU that previous the observed DMU;
• The DMU1 (the DMU with the lower quantity of inputs) consider the Eff$_{CCR \; O}$ equal to zero.

Algorithm:
Step 1: CALCULATE THE CCR EFFICIENCY of the analysed DMUs.

Step 2: CALCULATE , Δ_{Oq}^{ef} such tha $\Delta_{Oq}^{ef} = (Ef_{CCR\,O} - Ef_{CCR\,q})$, for all q.

Step 3: DEFINE INEFFICIENT THE DMUs that:, $\Delta_{Oq}^{ef} < 0$ for any q.
Step 4: DEFINE EFFICIENT THE DMUs that: $\Delta_{Oq}^{ef} > \theta$ for all q.

Example of FConc application:
This section will present numerical example to elucidate the understanding of FConc algorithm application. The CCR efficiency is calculated by SIAD software.

Table 1 Example of application of the FConc algorithm

DMU	INPUT	OUTPUT	CCR Efficiency (Steps 1,2)
1	0,5	0,5	0,321429
2	1,5	3	0,642857
3	2,5	6	0,771429
4	3,5	9	0,826531
5	4,5	14	1,000000
6	6	18	0,964286
7	7,5	21	0,900000
8	8	22	0,883929
9	8,5	22,5	0,850840

Steps 3 and 4: Calculate $\Delta_{m,q}^{ef}$ where is verified that:

- DMUs 1, 2, 3, 4, 5 are considered efficient.
- DMUs 6, 7, 8, 9 are considered inefficient.

4   Comparative Analysis of Results

The following figure 2, could represent hypothetic results of simulated models. The frontiers presented by the three presented models, with one input and one output, can be verified.

Following the basic presupposition of proportionality, the DEA CCR model just introduced one efficient DMU because no others have achieved the productivity of DMU more efficient. The CCR efficient was DMU7.

In models that assume variable returns to scale, it can be seen divergence in results. The model of the concave up frontier, FConc, ensures increasing returns to scale. The DMUs 1, 3, 4, 6 and 7 are considered efficient. They present constant increasing of productivity.

The DEA BCC model, with concave down frontier, presented only two efficient DMUs, DMU 1 and 7. However, the DEA BCC has, theoretically, the possibility of a boundary with increasing, decreasing and constant returns to scale. Although, in the figure 2, is exposed a case that the DMUs 3, 4, and 6 are not considered efficient.

Thus, the BCC border seems to neglect some efficient DMUs; what could be considered as a fragility since that the model proposes to check variable returns to scale.

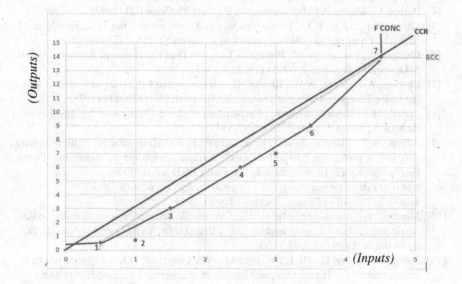

Fig. 2 Efficient Frontiers of (a) DEA CCR; (b) DEA BCC; (c) Fconc

5   Conclusion

The results of the study signalizes that the analysis of returns to scale of the DEA BCC model is limited by one specific kind. It is because, there are some cas-es that the BCC's boundary (concave down curve) is incapable to identify efficient DMUs with increasing returns to scale, as shown.

Although, the concave up border was able to identify the all the DMUs with no decreasing returns to scale; for that reason, that DMUs, should be considered efficient in a variable returns to scale model. This new model could be used as a complementary method for analyses variable returns to scale

As a future study, we suggest expanding the application of FConc, in order to compare with the results of BCC model to consolidate or not this fragility of the BCC model.

References

[1] Allencastro, L., Fochezatto, A.: Eficiência técnica na gestão de recursos em instituições privadas de ensino superior. Análise 17(2), 234–242 (2006)

[2] Banker, R.D., Charnes, A., Cooper, W.W.: Some Models for Estimating Technical and Scale Inefficiencies in Data Envelopment Analysis. Management Science 30(9), 1078–1092 (1984)

[3] Banker, R., Thrall, R.: Estimation of Returns to Scale Using Data Envelopment Analysis. European Journal of Operational Research 62, 74–84 (1992)

[4] Barbot, C., Castro, A.: Microeconomia, 2nd edn. McGraw-Hill (1997)

[5] Chang, K.P., Guh, Y.Y.: Linear Production Functions and the Data Envelopment Analysis. European Journal of Operational Research 52, 215–223 (1991)

[6] Charnes, A., Cooper, W.W.: Preface to Topics in Data Envelopment Analysis. Annals of Operations Research 2, 59–94 (1985)

[7] Charnes, A., Cooper, W.W., Rhodes, E.: Measuring the Efficiency of Decision Making Units. European Journal of Operational Research 2(6), 429–444 (1978)

[8] Farrell, M.J.: The measurement of productive efficiency. Journal of the Royal Statistic Society, Series A, parte 3, 253–290 (1957)

[9] Jubran, A.J.: Modelo de análise de eficiência na administração pública: estudo aplicado às prefeituras brasileiras usando a análise envoltória de dados. São Paulo: Escola Politécnica, Universidade de São Paulo, PHD thesis (2006).
`http://www.teses.usp.br/teses/disponiveis/3/3142/`
`tde-13122006-180402/` (access May 25, 2014)

[10] Martins, F., Soares de Mello, J.C.: Avaliação educacional aplicando análise envoltória de dados e apoio multicritério à decisão. In: XXIII Encontro Nac. de Eng. de Produção, Ouro Preto, MG (2003)

[11] Meng, W., Zhang, D., Qi, L., Wenbin Liu, W., Two-level, D.E.A.: approaches in research evaluation. The International Journal of Management Science 36(6) (2008)

[12] Panzar, J.C., Willing, R.D.: Economies of scale in multi-output Production. Quartely Journal of Economics 9, 481–494 (1977)

[13] Koopmans, T.: Activity analysis of production and allocation. John Wiley & Sons, NewYork (1951)

[14] Pareto, V.: Manuel d'Economie Politique. Giars &Briere, Paris (1909)

[15] Périco, A.E.: A relação entre as infraestruturas produtivas e o produto interno bruto (PIB) das regiões brasileiras: uma análise por envoltória de dados. São Carlos: Escola de Engenharia de São Carlos, Universidade de São Paulo.Tese de Doutorado em Engenharia de Produção (2009)

[16] Varian, H.: Microeconomic Analysis. W.W. Norton & Company, London (1992)

[17] Varian, H.: Microeconomia: Princípios Básicos. Campus, Rio de Janeiro (2000)

[18] Vasconcellos, M.A.S., et Oliveira, R.G.: Manual de Microeconomia. Manual de Microeconomia, São Paulo, Atlas, (2000). Dissertation Trent JW (1975) Experimental acute renal failure. Dissertation, University of California

Modeling Inter-sector Health Policy Options and Health Gains in a Long-term Care Network: A Location-Allocation Stochastic Planning Approach

Teresa Cardoso[1], Mónica Duarte Oliveira[1], Ana Paula Barbosa-Póvoa [1], and Stefan Nickel[2]

[1] Centre for Management Studies, Instituto Superior Técnico,
  Universidade de Lisboa, Lisbon, Portugal
[2] Institute of Operations Research, Karlsruhe Institute of Technology, Karlsruhe, Germany

Abstract. Although not a common practice, proper planning of health care networks should consider the attainment of health gains and the impact of inter-sector health policy options. This study proposes a multi-objective stochastic model to support location-allocation decisions in the long-term care sector that integrates these aspects. The model aims at maximizing expected health gains and minimizing expected costs, using the augmented ε-constraint method for analyzing this trade-off. The impact of inter-sector policy options on this trade-off is explored. A case study in Portugal is analyzed.

Keywords: OR in health services, LTC planning, Stochastic model, Health gains, Inter-sector policy options, Portugal.

1 Introduction

Long-term care (LTC) is a multidisciplinary approach aiming at improving the quality of life of individuals suffering from chronic illnesses and/or with disabilities (World Health Organization 2000). LTC comprises a wide range of services (including institutional, home-based and ambulatory services) and it can be provided by different entities (including the family, public and private entities).

LTC planning is currently faced as a key health policy priority in many European countries. The need for this planning is mainly related to the high (and increasing) levels of LTC demand together with the low provision of care found in these countries (World Health Organization 2000). In the current context of tight budget constraints, National Health Service (NHS)-based countries need to plan the delivery of LTC while minimizing costs and considering other health policy objectives, such as health gains and equity-related objectives.

© Springer International Publishing Switzerland 2015

A.P.F.D. Barbosa Póvoa and J.L. de Miranda (eds.), *Operations Research and Big Data*,
Studies in Big Data 15, DOI: 10.1007/978-3-319-24154-8_4

Literature in the area of health care planning shows an extensive use of location-allocation models based on mathematical programming (Brailsford and Vissers 2011). When planning networks of health care services in general, and LTC services in particular, one should consider: i) the pursue of multiple and often conflicting objectives (where health gains, cost and equity play a key role) (Smith et al. 2012; Mestre et al. 2015; Cardoso et al. 2016); ii) the multi-service nature of these services (Mestre et al. 2015); iii) the need for planning changes over time (Santibáñez et al. 2009; Mestre et al. 2015; Cardoso et al. 2016); iv) the uncertainties inherent to this sector (Mestre et al. 2015; Cardoso et al. 2013); and v) the impact of a variety of health policy options in planning decisions (Maenhout and Vanhoucke 2013). Although most of these aspects have been considered in existing literature in the area, few location-allocation studies have considered health gains and explored the impact of inter-sector health policy options in the design of networks of care. Moreover, no study exists comprehensively considering all these features in the health care sector in general, and in the LTC sector in particular. In fact, very few studies are devoted to the LTC sector (Lin et al. 2012; Cardoso et al. 2016).

This article aims to fill this gap in literature and proposes a multi-objective two-stage stochastic mixed integer linear programming (MILP) model to inform location-allocation decisions in the LTC sector in the context of a NHS-based country. A key model's feature is the introduction of health gains as a model objective. Together with health gains, costs and several equity concerns are also considered. While the maximization of expected health gains and the minimization of expected costs are accounted for as objectives, a variety of satisficing equity levels (equity of access, EA; equity of utilization, EU; socioeconomic equity, SE; and geographical equity, GE) are imposed as constraints. The augmented ε-constraint method is used to explore the trade-off between expected health gains and expected costs. An additional key contribution of this study is related to the modeling of inter-sector health policy options that may have impact on the health gains and cost trade-offs and that may influence the configuration of a network of care. The model also accounts for uncertainty in LTC demand and supply and for the multi-service nature of LTC. A case study in the Great Lisbon region in Portugal for the 2014-2016 period is explored, with planning decisions being evaluated on a yearly basis.

This article is organized as follows. Section 2 presents background information, and the methodology is detailed in Section 3, providing details on the mathematical programming model and on the selected solution's approach. Key results are explored in Section 4. Section 5 presents conclusions and future research.

2   LTC Planning Background

The location-allocation planning model developed in this study departs from a LTC network operating in the context of a NHS-based system. Such a system can be found, for example, in Portugal (Barros et al. 2011) and in England (Boyle 2011). For the purpose of this study, only the health care component of LTC is considered, with a wide range of institutional (IC; including short-term to longer institutionalizations), home-based (HBC) and ambulatory care (AC) services being modeled (Mission Unit for Long-Term Care 2013).

(Re)Organizing a LTC network involves making decisions on the services' location (when and where to deliver services) and capacity levels (how many beds and human resources should be made available per service and patients groups over time). Due to high pressures the healthcare sector is currently facing to control public spending, this (re)organization should account for the maximization of health benefits together with the minimization of costs. Since equity concerns are also a pillar of any NHS-based system, a variety of equity satisficing levels (namely, EA, EU, SE and GE) are accounted for in the form of model's constraints (Simon 1956).

The impact of adopting a variety of inter-sector health policy options in these planning decisions is also explored, namely: i) converting acute hospitals into LTC units; and ii) changing the LTC provision paradigm from an institutional towards a community-based paradigm (Rato et al. 2009). Different inter-sector policy strategies can be built by combining these different policy options. Furthermore, since it is not possible to foresee with total confidence how the number of individuals in need and length of stay (LOS) will evolve in coming years, the uncertainty on this information is also modeled.

3   Methodology

Considering the before-mentioned features of the planning problem, a two-stage stochastic MILP model is developed. First-stage decisions include the opening and closure of services, and second-stage decisions include decisions related to the allocation and reallocation of patients and resources and investments in extra bed capacity. In terms of constraints, the model should respect the following (for more details consult (Cardoso et al. 2013)):

- Opening and closure of services – opening/closing a service is not allowed after its closure/opening in a previous time period;
- Minimum level of demand satisfaction – a minimum level of satisfied demand per type of service is imposed over time;
- Single and closest assignment – individuals should receive the care they need in the closest available service and within a maximum travel time;
- Resources availability – the bed capacity delivered per service and location cannot assume values lower than the bed capacity required per scenario, which may result in excess capacity (similarly to (Birge and Louveaux 1997));
- Minimum and maximum capacity – a minimum and maximum number of beds and individuals in need are defined per service;

- Reallocation of beds – reallocating beds between services in different locations is restricted to the first time period, but is always permitted across services in the same location; and a maximum number of beds is allowed to be reallocated from each IC service;
- Equity satisficing levels – four satisficing levels of equity are used, namely, EA, EU, SE and GE, and are defined similarly to (Cardoso et al. 2013).These satisficing equity levels should be defined by decision-makers (DMs), and correspond to levels that should be accomplished (and not optimized) in the context of scarce resources (Simon 1956).

The two objectives considered in the model are the maximization of expected health gains and the minimization of expected costs. Costs cover both investment (related to investments in new beds and the reallocation of beds) and operational costs (related to the usage of beds and to HBC and AC delivery). Health gains are considered to depend on the type of service provided. To deal with these two objectives, the augmented ε-constraint method is used (Mavrotas 2009). When compared to the classical ε-constraint, this method avoids the production of weak Pareto optimal solutions. This is achieved by transforming the objective function constraint into an equality through the incorporation of a slack/surplus variable, with this variable being then used with a lower priority in the objective function and forcing the production of only efficient solutions (consult (Mavrotas 2009) for more details).

To deal with the uncertainty in the number of individuals in need and LOS, a scenario tree approach is used (Birge and Louveaux 1997). The number of different realizations considered for each uncertain parameter will limit the number of scenarios built.

4   Case-Study

This section starts by briefly describing the dataset used for the model application to the county level in the Great Lisbon region on a yearly basis over the 2014-2016 period, and then explores the results obtained.

4.1   Data-Set Used and Inter-sector Policies

The model was implemented in the General Algebraic Modeling System (GAMS) 23.7 and solved with CPLEX 12.0 on a Two Intel Xeon X5680, 3.33GHz computer with 12GB RAM. The dataset used is summarized in Table 1.

Quality-Adjusted Life Years (QALYs) gained with the delivery of each type of LTC service were used as a proxy for health gains, and were estimated using information on disabilities (Rato et al. 2009) together with the EQ-5D self-report questionnaire (Szende et al. 2007). In order to model uncertainty, the number of individuals in need (Cardoso et al. 2012) and the LOS (Central Administration of the Health System 2013) were taken as the uncertain parameters. The extended

Pearson-Tuckey method (Clemen and Reilly 2003) was employed to select three annual scenarios for each uncertain parameter, and these scenarios were then combined to build a scenario tree with 81 scenarios. In addition to this data, satisficing equity levels defined by the Head of the Long-Term Care Coordination Team of the Lisbon and Tagus Valley Regional Health Authority (who participated in this study in the role of a real DM in the LTC sector) were also used.

Table 1 Main information used for the model application.

Information	Source
LTC supply at the end of 2013 in the Great Lisbon region	(Mission Unit for Long-Term Care 2013)
Operational and investment costs	
Travel time (in minutes) between each county and each LTC service	(ViaMichelin 2014)
Maximum travel time allowed for LTC patients accessing IC services	(Health Regulation Authority 2011)
Information about physical, functional, mental and social disabilities of patients receiving care within the National Network of LTC (*Rede Nacional de Cuidados Continuados Integrados*, RNCCI) during 2008, before and after receiving LTC	(Rato et al. 2009)
The number of individuals requiring IC, HBC and AC	(Cardoso et al. 2012)
LOS associated per IC service	(Central Administration of the Health System 2013)

In order to analyze the influence of different inter-sector health policy options, three policy strategies are selected for analysis: Strategy I corresponds to the base strategy and is characterized by no hospital conversion (i.e., converting acute hospitals into LTC units is not allowed) and the usage of an institutional-based paradigm (i.e., it is not possible to substitute IC by HBC); Strategy II differs by having a community-based paradigm (with 50% of those needing IC receiving HBC), maintaining no hospital conversion; and Strategy III differs by allowing for hospital conversion, keeping the usage of the institutional-based paradigm.

4.2 Results

The Pareto Frontiers obtained when applying the multi-objective stochastic model to the county level in the Great Lisbon region in the 2014-2016 period under different inter-sector policy strategies are shown in Fig 1. These results are obtained when departing from the current network of LTC services that is characterized by a low provision of IC and HBC (with approximately 90% of unsatisfied demand (Cardoso et al. 2012)), as well as by a total lack of AC provision. These results allow taking conclusions about how the reorganization of the LTC network under different policy strategies impact on the achievement of health gains and at what cost. For all the frontiers, solution A is characterized by the lowest level of LTC provision (as traduced in the minimum expected cost and health gains), whereas solution K is characterized by full LTC provision (i.e., demand for IC, HBC and

AC is fully satisfied, thus resulting in the highest expected cost and health gains). Each solution represented in Fig. 1 was obtained by imposing a 180 minutes limit for the computation time, with optimality gaps below 0.5% for all the solutions.

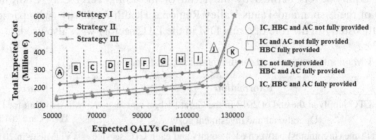

Fig. 1 Pareto Frontiers obtained when applying the multi-objective stochastic model under different inter-sector policy strategies. Legend: IC – Institutional Care; HBC – Home-Based Care; AC – Ambulatory Care.

Fig. 1 shows similar shapes of the Pareto frontier across strategies. In particular, taking Strategy I as example: moving from solution A to solution B implies extra investments to achieve full provision of HBC; moving from solution B towards solution I demands for extra investments in AC provision, improving unsatisfied demand from 59% to 7%, respectively; and moving from solution I to solution J is associated with investments in both AC and IC, with full provision of AC achieved in solution J, whereas unsatisfied demand for IC improves from 29% to 20%.

Fig. 1 also shows that reorganizing the LTC network under Strategy II allows achieving the highest levels of QALYs gained, when compared to the remaining strategies and for the same level of cost. This is mainly due to the substitution of IC by cheaper HBC provision. Also, a sharp increase on costs is found as one moves from solution J to K, mainly because investments fully target IC provision, which is more expensive than HBC or AC. Furthermore, when compared to Strategy I, Strategy III also leads to higher QALYs gained for the same level of cost. This is due to the use of bed capacity installed in existing acute hospitals, thus avoiding extra investments for this bed capacity. Similar conclusions can be taken from Table 2, where it can be seen that Strategy III is always more cost-effective than Strategy I. Table 2 also confirms that Strategy II is the most cost-effective strategy – the additional cost per QALY gained is the lowest one across solutions.

Table 2 Additional cost (in thousands of euros) per QALY gained (Incremental Cost Effectiveness Ratios, ICERs) with reference to the current LTC network

Strategy	Solution										
	A	B	C	D	E	F	G	H	I	J	K
I	2.8	2.6	2.4	2.3	2.2	2.1	2.0	1.9	1.8	1.9	4.2
II	1.0	1.1	1.1	1.1	1.2	1.2	1.3	1.3	1.3	1.3	2.0
III	1.9	1.8	1.7	1.7	1.7	1.6	1.6	1.6	1.5	1.6	4.0

Fig. 1 also shows that an inadequate supply of LTC currently exists in the Great Lisbon region – in fact, improving the delivery of LTC so as to achieve a full provision of care will cost a minimum of 350 million euros for the three-years period (if Strategy II is adopted). That amount is significantly higher than the budget currently available for reorganizing and operating the LTC network in that area (corresponding to 90 million euros).

5   Conclusion

Planning networks of LTC is a top policy priority across European countries, not only as it targets the delivery of care to those that still need it, but also as it contributes for lowering the costs of the healthcare system as a whole. This planning is even more relevant for NHS-based countries facing strong pressures to control their public health care spending.

This study proposes a multi-objective two-stage stochastic MILP model to assist location-allocation decisions in the LTC sector in NHS-based countries. Expected costs and health gains-related objectives are accounted for in the model, whereas equity satisficing levels are modeled as constraints. The model also allows exploring the impact of a variety of inter-sector health policy options.

Key results obtained from applying the model to the Great Lisbon region in Portugal for the 2014-2016 period on a yearly basis show that an inadequate provision of LTC exists in the region. The model results have also shown the relevance of exploring the impact of adopting different inter-sector health policy options, given that the trade-off between expected cost and health gains has shown to be sensitive to these policies.

Future work may consider exploring the sensitivity of the model results to changes in the number of scenarios in use. An additional line of research includes the use of Multiple Criteria Decision Making methods so as to assist decision makers selecting the most preferred solution, when departing from solutions generated using the augmented ε-constraint method. In addition, accounting for equity concerns as objectives, together with health gains and costs, should also be explored.

Acknowledgments. The first author acknowledges financing from the *Fundação para a Ciência e Tecnologia* (Portugal) (SFRH/BPD/98270/2013). The authors thank the support from the Institute of Operations Research, Karlsruhe Institute of Technology (KIT, Germany) and from the Centre for Management Studies of *Instituto Superior Técnico* (CEG-IST, Portugal). The authors thank Dr. Regina Sequeira Carlos, the Head of the Long-Term Care Coordination Team of the Lisbon and Tagus Valley Health Authority, for her willingness to take part in this study. The authors also gratefully acknowledge the helpful suggestions of the reviewers.

References

Barros, P.P., Machado, S.R., Simões, J.A.: Portugal: Health system review. Health Systems in Transition 13(4), 1–156 (2011)

Birge, J.R., Louveaux, F.: Introduction to stochastic programming. Springer, New York (1997)

Boyle, S.: United Kingdom (England): Health system review. Health Systems in Transition 13(1), 1–483 (2011)

Brailsford, S., Vissers, J.: OR in healthcare: A European perspective. Eur. J. Oper. Res. 212(2), 223–234 (2011)

Cardoso, T., Oliveira, M., Barbosa-Póvoa, A., Nickel, S.: Modeling the demand for long-term care services under uncertain information. Health Care Manag. Sci. 15(4), 385–412 (2012)

Cardoso, T., Oliveira, M., Barbosa-Póvoa, A., Nickel, S.: Moving towards an equitable long-term care network: A multi-objective and multi-period planning approach. Omega 58, 69–85 (2016)

Cardoso, T., Oliveira, M.D., Barbosa-Póvoa, A., Nickel, S.: Planning a long-term care network with uncertainty, strategic policy and equity considerations: A stochastic planning approach. Working Paper CEG-IST no. 4/2013, Instituto Superior Técnico (2013)

Central Administration of the Health System, Implementação e Monitorização da Rede Nacional de Cuidados Continuados Integrados - Relatório Final [Implementing and Monitoring the National Network of Long-Term Care - Final Report]. Cuidados Continuados - Saúde e Apoio Social (2013)

Clemen, R.T., Reilly, T.: Making Hard Decisions with Decision Tools Suite Update, Duxbury (2003)

Health Regulation Authority, Estudo do acesso dos utentes aos cuidados continuados de saúde [Study on patients' access to long-term care]. Oporto (2011)

Lin, F., Kong, N., Lawley, M.: Capacity planning for publicly funded community based long-term care services. In: Johnson, M.P. (ed.) Community-Based Operations Research. International Series in Operations Research & Management Science, vol. 167, pp. 297–315. Springer, New York (2012)

Maenhout, B., Vanhoucke, M.: An integrated nurse staffing and scheduling analysis for longer-term nursing staff allocation problems. Omega 41(2), 485–499 (2013)

Mavrotas, G.: Effective implementation of the e-constraint method in Multi-Objective Mathematical Programming problems. Appl. Math. & Comput. 213(2), 455–465 (2009)

Mestre, A.M., Oliveira, M.D., Barbosa-Póvoa, A.P.: Location–allocation approaches for hospital network planning under uncertainty. Eur. J. Oper. Res. 240, 791–806 (2015)

Mission Unit for Long-Term Care, Unidade de Missão para os Cuidados Continuados Integrados [Mission Unit for Long-Term Care] (2013), http://www.rncci.min-saude.pt (accessed September 2013)

Rato, H., Rodrigues, M., Rando, B.: Estudo de caracterização dos utentes da Rede Nacional de Cuidados Continuados Integrados – Relatório Final [Study characterizing the patients from the National Network of Long-Term Care - Final Report]. INA, Oeiras (2009)

Santibáñez, P., Bekiou, G., Yip, K.: Fraser Health uses mathematical programming to plan its inpatient hospital network. Interfaces 39(3), 196–208 (2009)

Simon, H.A.: Rational Choice and the Structure of the Environment. Psychol. Rev. 63(2), 129–138 (1956)

Smith, H.K., Harper, P.R., Potts, C.N.: Bicriteria efficiency/equity hierarchical location models for public service application. J. Oper. Res. Soc. 64(4), 500–512 (2012)

Szende, A., Oppe, M., Devlin, N.: EQ-5D value sets: inventory, comparative review and user guide. EuroQol Group MonographsThe, vol. 2. Springer, The Netherlands (2007)

ViaMichelin, ViaMichelin Maps & Route Planner (2014). http://www.viamichelin.co.uk/ (accessed September 2014)

World Health Organization. Home-based long-term care: report of a WHO study group. Technical Report Series, vol. 898. World Health Organization, Geneva (2000)

Optimization of Construction and Demolition Waste Management: Application to the Lisbon Metropolitan Area

Mafalda Neiva Correia[1], Marta Castilho Gomes[1,*], and Joaquim Duque[2]

[1] CERIS-CESUR, Instituto Superior Técnico, Universidade de Lisboa,
  Av. Rovisco Pais, 1049-001 Lisboa, Portugal
[2] UAER, Laboratório Nacional de Energia e Geologia,
  Estr. do Paço do Lumiar, 22, 1649-038 Lisboa, Portugal
  marta.gomes@tecnico.ulisboa.pt

Abstract. This work presents a mixed integer linear programming (MILP) model of a construction and demolition waste (CDW) recycling network that includes a methodology for environmental impact assessment (Eco-indicator 99). Applied to a geographical area, this generic formulation can be used for feasibility studies and sensitivity analyses of the input parameters, while simultaneously defining the optimal network design, *i.e.*, the location and capacity of the installed facilities and material flows. Furthermore, by using the ε-constraint methodology, the Pareto Front is drawn, allowing the decision maker to choose the solutions presenting the best tradeoff between economic and environmental goals.

The model was applied to a Lisbon Metropolitan Area case study, which required extensive data collection and estimation. Different instances of the problem were generated, by varying conditions and model parameters (sensitivity analysis), and the MILP model solved. Solutions comprise a small number of high capacity recycling plants, when cost minimization is at stake, while when preference is given to the minimization of environmental impacts solutions comprise several geographically dispersed recycling plants of low capacity.

1 Introduction

The European Union (EU) has not yet published any specific legislation for the management of CDW. However, in Portugal, its management is subject to specific regulations since 2008. Currently, the legislation that is in force is the Decree-Law n° 73/2011 which establishes the legal regime on waste management. Despite the existing legislation, Portugal is far from achieving the 2020 EU target, which stipulates that all Member States shall take the necessary measures to ensure that at least 70% (by weight) of non-hazardous CDW are prepared for reuse, recycling

---

* Corresponding author.

© Springer International Publishing Switzerland 2015
A.P.F.D. Barbosa Póvoa and J.L. de Miranda (eds.), *Operations Research and Big Data*,
Studies in Big Data 15, DOI: 10.1007/978-3-319-24154-8_5

and recovery. At national level, the majority of CDW produced is still landfilled and it is therefore important to outline a strategy for its management. The construction industry is one of the most active sectors across Europe, representing 28.5% of the employment in industry (Mália et al., 2011). This activity has led to serious consequences on the environment as well as management problems. In 2008, the annual production of CDW in EU was, approximately, 100 million tons (Decree-Law nº46/2008). As in the EU, a very significant part of the waste produced in Portugal has its origin in the construction industry. In what concerns the destination of CDW, about 76% of CDW produced in Portugal is landfilled, 11% is reused, 4% incinerated and only remains around 9% that is actually recycled (Coelho and Brito, 2011).

Although efforts have been made in the development of decision models to solve problems involving supply chains of residues, there are few papers that focus on planning and modeling of CDW chains. Moreover, the majority of works that have been published in this context use approaches such as systemic mapping techniques, dynamical systems and discrete simulation models (Yuan and Shen, 2011).

In this research an optimization model for planning a CDW recycling network was developed following the one presented by Hiete *et al.* (2011). The environmental perspective was subsequently embedded in the model following the methodology for environmental impact assessment developed by Duque (2007). Based on an extensive process of data collection, a case study of Lisbon Metropolitan Area was developed, involving the generation of several model solutions. The paper concludes with the corresponding results analysis and discussion.

2 Optimization Model for Planning a CDW Recycling Network

The optimization model takes three types of materials into account:

- Construction and Demolition Waste (CDW);
- Processed Materials (PM), which result from transformation of CDW in recycling plants;
- Residual Material (RM) originated by CDW processing in recycling plants.

The management options for CDW comprise the direct disposal at landfills or the processing in recycling plants with different levels of performance, leading to different shares of PM and RM. Concerning these materials, PM is sold and RM is sent for disposal in landfills. Moreover, the model does not impose total satisfaction of PM demand.

The mixed-integer linear programming (MILP) model developed considers:

- network nodes characterized by CDW production, PM demand, number and type of recycling plants, existing landfills;

- network arcs defined for all pairs of nodes (with an associated distance);
- transportation cost (as a function of distance);
- landfilling cost of CDW and RM and sales value of PM;
- different types of recycling plants (which differ in transformation characteristics and capacity). Besides the existing recycling plants (to be kept), new plants can be built but only one recycling plant of each type is allowed in each node;
- capacity, investment cost (only considered for new recycling plants) and variable costs for each type of recycling plant;
- interest rate and planning horizon;
- no limit for total capacity of landfills is assumed.

Decision variables in the model regard:

- opening of new recycling plants (where location and type of new recycling plants are defined);
- amount of materials transported between each pair of nodes;
- amount of material (CDW) processed in a recycling plant of a given type at each node;
- amount of materials landfilled (CDW or RM) or sold (PM) at each node;
- variables related to environmental impact assessment.

The first type of variables are binary while all others are positive (and continuous). The objective function, to be minimized, is the total cost of the CDW recycling network and includes: transportation costs between nodes (for the three types of materials); landfilling costs for CDW and RM and revenues from PM sale; costs of CDW processing in recycling plants; investment cost of the recycling plants to be built.

The Eco-indicator method, published in 1999, was developed by a group of researchers from the Dutch company PRé-Consultants (PRé-Consultants, 2001). The Eco-indicator of a material or process is represented by a score that reflects the environmental impact, based on data from life cycle assessment. The definition of this indicator takes into account three types of damage: damage to human health, damage to ecosystems quality and damage to natural resources. The EI99 methodology was embedded in the model constraints.

The complete set of constraints is described as follows:

- The existing recycling plants are not shut down.
- Regarding the transformation at recycling plants, mass balances of materials are taken into account. These equations have different forms for CDW (input material) and for PM and RM (output materials).
- Maximum demands of PM as well as maximum amounts of CDW and RM sent to landfills are imposed.
- Capacity constraints set a limit on the amount of CDW that can be processed at recycling plants.
- To assess the environmental impact, all elements in the inventory are accounted for: pollutants emitted by heavy-duty vehicles that transport materials between

nodes, total emissions of pollutants due to electricity consumption at recycling plants, fuel consumption by heavy-duty vehicles and land use (by recycling plants).

- The amount of each type of damage, the total amount of damages and the EI99 value are computed.
- A maximum limit for the EI99 value is established (implementation of the ε-constraint methodology).

3   Data Collection

Developing a systemic model for a regional network for CDW collection, transport, recovery and reuse requires in its first phase estimating CDW generation in the considered region. Bernardo (2013) devised an innovative process of CDW estimation that: (i) collects and processes data of real demolition works, (ii) characterizes CDW generation by building type and (iii) extrapolates the CDW indicators for the different geographical sub-areas of the studied area. The generality of this methodology allows it to be applied to any geographical area.

The procedure was illustrated for the Lisbon Metropolitan Area (LMA) using data from around 50 demolition works between 2008 and 2012, collected from *Ambisider*, one of the market leaders of the Portuguese demolition sector. From this, and based on the existing statistical information, CDW indicators were estimated both at the municipal and the parish levels (LMA comprises 18 municipalities and 211 parishes). The total amount of CDW generated in LMA in 2012 was estimated to be 821 kton.

To implement the MILP model, and according to Coelho (2012), CDW was divided into two types of materials: 41% of "Concrete" and 59% of "Other CDW". Two types of processed materials (PM) were considered - high-quality recycled material (HQRC) and low-quality recycled material (LQRC) – and a residual material (RM) produced by CDW processing in recycling plants. The demand of HQRC and LQRC by municipality (the geographical level considered in the present work), was obtained based on the shares of resident population. Furthermore, the landfills and recycling plants (RP) that exist in LMA were considered in the model: there are 3 licensed landfills (one in Lisbon municipality and two in Setúbal) and 2 RP in the municipalities of Sintra and Seixal (*Agência Portuguesa do Ambiente*). The data needed to characterize the RPs was gathered by live query; the existing RP are of low-quality, low-capacity type (LQLC). The landfilling costs of CDW and RM and the sales value of PM were considered uniform for the study area. The landfilling costs of "Concrete" and "Other CDW" as well as the sales value of the PMs were determined based on Coelho and Brito (2011). The landfilling cost of RM was defined according to Coelho (2012).

Values needed to implement the EI99 methodology were obtained from the EI99 manuals, available online. Finally, an interest rate of 5% and a planning horizon of 10 years were considered.

4 Case Study Results

The MILP model was implemented in GAMS modeling language and solved with CPLEX 12.5.0.0 in an Intel® Core™ i7 CPU, Q720, with 1.6GHz and 6 GB of RAM. The default upper bound for the optimality gap was considered; CPU time for solving the model is less than 1 sec.

In the first stage of the experimental study, the Pareto Front was obtained using the ε-constraint method, one of the approaches for solving multi-objective problems (Barrico, 2007). First, anchor points for cost and EI99 minimization were computed (solutions 1 and 2 in Fig. 1, respectively). Then, cost was minimized while imposing an upper bound on the EI99 value, which was gradually reduced: solutions 1.1 to 1.9 were thus obtained. The resulting curve shows that higher cost solutions present lower environmental impacts. Besides solutions 1 and 2, solutions 1.5 and 1.9, also highlighted in Fig. 1, were selected and kept for further analysis (appearing in the charts of Fig. 2 and 3).

Solution 1.5 was considered noteworthy because, for a cost increase of 1.5%, compared with solution 1, a reduction of 31% in the EI99 value was achieved. On the other hand, solution 1.9 presents a reduction of 57% in the EI99 value compared to solution 1 (the EI99 value being very close to the one in solution 2) but a cost increase of 29%, significantly smaller than the one in solution 2. Solutions 1 and 1.5 present 3 RP, solution 1.9 has 7 RP and solution 2 presents 8 RP.

In the second stage, the MILP model was solved for cost minimization (with no constraint on the EI99 value) under different conditions, judged interesting to assess, which includes a sensitivity analysis to parameter values. These solutions are thus related to solution 1 above and can be outlined as follows.

Solutions:
3   No pre-existing recycling plants in LMA
A   Decrease of disposal costs of RM
B   Use of the original transformation matrix in Hiete *et al.* (2011)
C-R Sensitivity analysis to parameter values (one at a time), considering positive and negative variations of 20%.

Fig. 2 and 3 summarize the results. Fig. 2(a) represents the final destination of CDW (in percentage of the total amount of CDW): direct disposal in landfills; disposal as RM from the recycling process or conversion into PM (either HQRC or LQRC). Fig. 2(b) represents the cost share, *i.e.* the contribution of fixed costs, transportation costs, disposal costs and variable costs to the total cost. Finally, solution environmental impact is depicted in Fig. 3, comprising damage to human health, damage to ecosystem quality and damage to mineral and fossil resources.

Regarding the number of recycling plants, solution 3 presents 1 RP, solution A has 4 RP and solutions B-R have 3 RP. Solution analysis shows that, for all solutions with 3 RP, the new recycling plant corresponds to a HQHC (high quality, high capacity) plant located in the municipality of Lisbon.

Fig. 2(a) shows solutions 2 and 3 to be the ones with the highest direct disposal rate (74%). Fig 2(b) underlines that disposal costs are the ones contributing most to total cost, with an average share of 66%. Fixed costs are related to the number of recycling plants installed and thus this proportion is higher for solution 2. Fig. 3 shows the larger environmental impacts to be the ones on mineral and fossil resources as well as on human health.

The following general conclusion can be drawn from the analysis undertaken:

- Cost minimization requires the reduction of landfilled material, giving rise to the installation of a low number of high capacity recycling plants.
- Minimization of environmental impact leads to the installation of several geographically dispersed, low capacity recycling plants so as to minimize transportation costs.

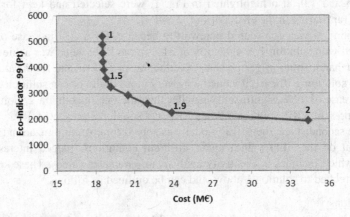

Fig. 1 Pareto Front (with noteworthy solutions).

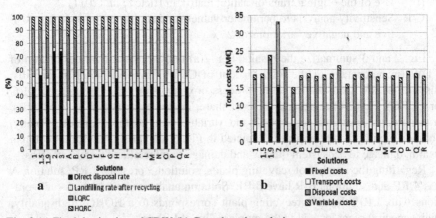

Fig. 2 (a) Final destination of CDW (b) Cost share by solution.

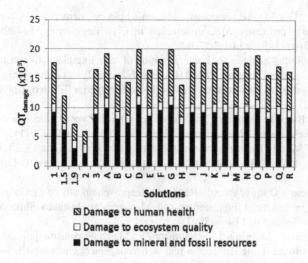

Fig. 3 Damages to human health, ecosystem quality and to mineral and fossil resources.

5   Conclusion and Future Developments

This work proposes an MILP model for managing CDW in a region, which offers an interesting potential since it combines a strategic planning component (associated to definition of the number, location and capacity of recycling plants to install) with a lower level tactical component (associated to the definition of material flows in the network). Furthermore, a significant contribution of this work regards consideration of environmental aspects in the model through implementation of the EI99 methodology.

For future work it would be interesting to compare the present results with those considering satisfaction of demand at each node. Application of the EI99 methodology only took into account damages caused by emissions of heavy vehicles, fuel consumption, electricity consumption and land use of recycling plants. Therefore, land use and emissions of pollutants from landfills should also be considered when computing damages. Finally, further refinement of the CDW network is desirable, namely regarding operations in recycling plants.

References

Barrico, C.: Evolutionary multi-objective optimization in uncertain environments - research for robust solutions. PhD thesis in Electrical and Computer Engineering, Faculdade de Ciências e Tecnologia, Universidade de Coimbra (2007)

Bernardo, M.: Gestão dos resíduos de construção e demolição: Caracterização, quantificação e processos. MSc dissertation in Civil Engineering, Instituto Superior Técnico, Universidade de Lisboa (2013)

Coelho, A.: Feasibility analysis for implementation of recycling plants of construction and demolition waste in Portugal - PART III - Feasibility analysis of a recycling plant. Report in the context of post-doctoral studies. Instituto Superior Técnico, Universidade de Lisboa (2012)

Coelho, A., de Brito, J.: Economic analysis of conventional versus selective demolition - case study. Resources, Conservation and Recycling 55(3), 382–392 (2011)

Decree-Law n°73/2011. The management system of waste. Lisbon, June 17, 2011

Decree-Law Nr. 46/2008. The system of construction and demolition waste, Lisbon, March 12, 2008

Duque, J.: Síntese e Optimização de Redes de Reaproveitamento de Produtos Residuais. PhD thesis in Industrial Engineering and Management, Instituto Superior Técnico, Universidade Técnica de Lisboa(2007)

Hiete, M., Stengel, J., Ludwing, J., Schultmann, F.: Matching construction and demolition waste supply to recycling demand: a regional management chain model. Building Research & Information 39(4), 333–351 (2011)

EEA (European Environment Agency). EMEP/EEA air pollutant emission inventory guidebook - 2009. Technical report n°9. Chapter 1.A.3.b Road transport (2009) (update May 2012)

Mália, M., Brito, J., de, B.M.: Indicators of construction and demolition waste for new residential buildings. Ambiente Construído, Porto Alegre 11(3), 117–130 (2011)

PRé Consultants, The Eco-indicator 99 - A damage oriented method for Life Cycle Impact Assessment. Methodology Report (2001)

Yuan, H., Shen, L.: Trend of the research on construction and demolition waste management. Waste Management 31(4), 670–679 (2011)

Financial Structure, Product Market Decisions and Default Risk in an Asymmetric Duopoly

Magali Costa[1] and Cesaltina Pacheco Pires[2]

[1] ESTG, CEFAGE-UE and Instituto Politécnico de Leiria, Portugal
[2] CEFAGE-UE and Departamento de Gestão, Universidade de Évora, Portugal

Abstract. Financial and output market decisions are crucial to the success or failure of an organization. In this paper we analyze the equilibrium default risk in a two-stage duopoly model with an uncertain environment, where firms decide their financial structure in the first stage of the game and decide their quantities in the second stage of the game. Using numerical analysis, we analyze the impact of changing the asymmetry in the two firms' marginal costs on the equilibrium default risk. Our results show that as a firm becomes less efficient it is optimal to reduce its debt level and the quantity produced. The reverse is true for the more efficient firm. This behavior implies that although higher marginal cost leads to lower profits, the less efficient firm reduces its default probability due to a more cautious behavior in the financial and product market.

Keywords: Capital structure, Product market competition, Default risk JEL classification, D43, G32, G33.

1 Introduction

Over the last decades, the financial literature has addressed the issue of default and bankruptcy risk. Considering its negative social and economic impact on the economy, it is not surprising that the existing literature has focused mainly on the best form to predict default and bankruptcy risk (for a survey of the empirical literature see Balcaen and Ooghe, 2006). However, in spite of the vast empirical literature, there is a lack of theoretical models aimed at understanding the factors that influence the default probability. The main objective of this paper is to provide a contribution in this direction.

The paper examines analytically and numerically, how the market structure influences financial decisions and product market decisions and, consequently, the default risk. Our objective is to study the impact of changes in the asymmetry between the firms' marginal production costs on the equilibrium default risk.

Brander and Lewis (1986) were the first to examine the relationship between financial decisions and output market competition. They consider a two stage Cournot duopoly model with an uncertain environment. In the first stage, each

© Springer International Publishing Switzerland 2015

A.P.F.D. Barbosa Póvoa and J.L. de Miranda (eds.), *Operations Research and Big Data*,
Studies in Big Data 15, DOI: 10.1007/978-3-319-24154-8_6

firm decides the capital structure. In the second stage, taking into account their previously chosen financial structure, firms take their decisions in the output market. The model focuses on the effects of the limited liability in debt financing. The authors conclude that debt tends to encourage a more aggressive behavior by the indebted firm in the output market, while the competitor tends to produce less.

The relationship between the financial structure decisions, the output market decisions and the default probability has been analyzed theoretically by a small number of authors. Franck and Le Pape (2008) and Haan and Toolsema (2008) are among these few authors. While Brander and Lewis (1986) present a general model, without specifying whether products are homogenous or differentiated and whether uncertainty affects demand or costs, Franck and Le Pape (2008) and Haan and Toolsema (2008) have explored more specific models and used numerical simulations to analyze the impact of demand uncertainty and the degree of product substitutability on the probability of default. The authors come to similar conclusions: the probability of default is decreasing with the degree of product substitutability when goods are complementary and it is increasing with the degree of product substitutability when the goods are substitutes. Moreover, the default probability is decreasing with the level of uncertainty. Although these results are quite interesting, they ignore the possibility of firms having different degrees of efficiency. In this paper, our aim is to extend their results by incorporating this important aspect of reality.

The remainder of the paper is organized as follows. In the next section we present the Model. Next we analyze the second stage of the game and the subgame perfect equilibrium. The following section presents the results while the last section summarizes the main conclusions of the paper.

2   Model

This study considers a particular case of Brander and Lewis (1986) model, where the duopolists produce differentiated products, demand is linear, marginal costs are constant and the uncertainty in the model is on the demand side. A similar model has been considered by Haan and Toolsema (2008). The difference is that we do not assume that firms have the same marginal cost.

In the first stage each firm (firm i and firm j) decides the financial structure, i.e., the level of debt and equity in the capital structure. In the second stage each firm chooses the quantity to produce. The inverse demand function is given by:

$$p_i = \alpha + z_i - q_i - \gamma q_j \tag{1}$$

Where q_i and q_j are the quantities consumed of firm i and firm j products, respectively and p_i and p_j are the corresponding prices. The parameter γ, with $\gamma \in [0,1]$, corresponds to the degree of substitutability between the two firms products. Parameter α represents the expected size of the market. The observed size of the market depends on the random variable z_i that represents the effect of an exogenous demand shock, in other words, there is uncertainty regarding the size of the market. It is assumed that this variable is distributed in the interval $[-z_{max}, z_{max}]$, according to the uniform density function. We assume that z_i and z_j are independent and identically distributed.

In the first stage, firms choose simultaneously their debt levels so as to maximize the value of the firm. The value of the firm is equal to the sum of the equity value and debt value. We represent the debt obligation of firm i by b_i. b_i is the amount that the firm i pays at the end of the game to bondholders, if it has sufficient operating profits to do so. If the realized operating profits are less than b_i, all the operating profits obtained will be used to pay bondholders. The operating profits (revenue less costs, where c_i represents the marginal cost of production) are given by:

$$R_i = (\alpha - q_i - \gamma q_j + z_i - c_i)q_i \tag{2}$$

In the second stage of the game the manager maximizes the expected equity value, which is given by:

$$V^i(q_i, q_j, b_i, z_{max}) = \int_{z_{c_i}}^{z_{max}} ((\alpha - q_i - \gamma q_j + z_i - c_i)q_i - b_i)\frac{1}{2z_{max}}dz_i \tag{3}$$

Where $z_{ci}(q_i, q_j, b_i)$ is the critical value of z_i such that the operating profit of the firm is just enough for the firm to meet its debt obligations. This critical state of the world is implicitly defined by:

$$(\alpha - q_i - \gamma q_j + z_i - c_i)q_i - b_i = 0 \tag{4}$$

It should be noted that the critical state of the world is influenced by the quantity choices of the two firms and by the firm's debt level. This implies that $z_{ci}(q_i, q_j, b_i)$ is determined endogenously in the second stage of the game.

In the first stage firms choose simultaneously their debt levels so as to maximize the value of the firm. The value of the firm is equal to the sum of the equity value and debt value. The expected value of debt is given by:

$$W_i(q_i, q_j, b_i, z_{max}) = \Pr(z_i > z_{c_i})b_i + \int_{-z_{max}}^{z_{c_i}} (\alpha - q_i - \gamma q_j + z_i - c_i)q_i \frac{1}{2z_{max}}dz_i. \tag{5}$$

Note that b_i is different from W_i. b_i is the amount that firm i promised to pay at the end of the game to bondholders. W_i is the expected value of debt, which takes into account the probability of the firm not paying in full b_i, i.e. if this probability is positive $W_i < b_i$

Thus, the value of the firm is equal to the expected operating profits of the firm:

$$Y^i(q_i, q_j, b_i, z_{max}) = \int_{-z_{max}}^{z_{max}} (\alpha - q_i - \gamma q_j + z_i - c_i)q_i \frac{1}{2z_{max}}dz_i \tag{6}$$

3   Solving the Model

To solve this two stage dynamic game, we need to use the concept of subgame perfect Nash equilibrium (SPNE). The game is solved backwards, that is, one starts by determining the Nash equilibrium in the second stage of the game as a function of the debt levels chosen by the firms in the first stage. Then we solve the first stage game when firms make their financial decisions, so as to maximize the value of the firm.

In the second stage of the game, firm i chooses q_i, so as to maximize the expected equity value. Using Leibniz rule, the first-order condition of this problem is:

$$\int_{z_{c_i}}^{z_{max}} (\alpha - 2q_i - \gamma q_j + z_i - c_i)\frac{1}{2z_{max}}dz_i - [(\alpha - 2q_i - \gamma q_j + z_{c_i} - c_i)q_i - b_i]\frac{1}{2z_{max}}\frac{\partial z_{ci}}{\partial q_i} = 0 \qquad (7)$$

However, taking into account the definition of z_{ci}, the second term is equal to zero. Thus, after integrating the first term, the first-order condition is given by:

$$z_{max} + 2\alpha - 4q_i + z_{c_i} - 2\gamma q_j - 2c_i = 0 \qquad (8)$$

The first order condition for firm j is derived in a similar manner. Note that the first order conditions depend on the critical states of the world, z_{ci} and z_{cj}, which in turn depend on q_i and q_j. This implies that, in order to get the Nash equilibrium of the second stage game, we need to solve simultaneously the system of the two first order conditions and the two conditions that define the critical states of the world. This 4 equations system is equivalent to solving a polynomial equation of the fourth order, which does not have a simple analytical solution.

We developed a GAUSS code to solve the model numerically. Considering the various types of possible equilibria, we ran simulations for many values of the parameters γ and z_{max}. For each set of parameter values, we determine the Nash equilibrium of the second stage game, for many possible combinations of the debt levels and then for each (b_i, b_j) the equilibrium value of each firm (Y_i, Y_j) is computed. The Nash equilibrium of the debt game occurs when we find a vector (b_i^{**}, b_j^{**}) such that the two firms are simultaneously in their best responses. Thus (b_i^{**}, b_j^{**}) denotes the SPNE levels of debt. Finally, considering (b_i^{**}, b_j^{**}) the corresponding SPNE quantities (q_i^{**}, q_j^{**}) are computed as well as other equilibrium variables like the default probabilities $(\theta_i^{**}, \theta_j^{**})$. The default probability is given by:

$$\theta_i^{**} = \frac{z_{c_i}^{**} + z_{max}}{2z_{max}} \qquad (9)$$

4   Results

In this section we analyze how the equilibrium changes with a unilateral increase in the marginal cost of firm j for several values of the uncertainty level, z_{max}, and degree of product substitutability, γ. We consider that firm i has a null production cost, $c_i=0$, while firm j has marginal cost c_j. We examine how the equilibrium values of debt, output and default probabilities vary as c_j changes between 0 and 0.5 (c_j is represented in the x-axis), considering three possible values for the degree of product substitutability, γ ($\gamma=0.2$, $\gamma=0.6$ and $\gamma=1$). Three graphs are presented for each variable (the first corresponds to $z_{max} = 0.85$, the second to $z_{max} = 1.25$ and the third one to $z_{max} = 1.85$).

Figures 2 and 3 show the debt levels of firm i and firm j, respectively, as a function of the marginal cost of firm j. They allow us to conclude the following:

Result 1. *The SPNE level of debt of firm i, b_i^{**}, is increasing with the marginal costs of firm j, c_j. The increase is higher for high levels of demand uncertainty. On the contrary, the SPNE level of debt of firm j, b_j^{**}, is decreasing with the marginal cost of firm j, c_j. The decrease is larger for high levels of demand uncertainty.*

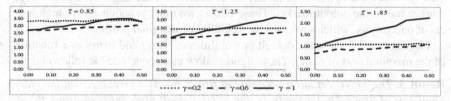

Fig. 1 SPNE debt level of the efficient firm as a function of the marginal costs of the rival

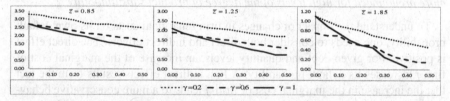

Fig. 2 SPNE debt level of the inefficient firm as a function of its marginal costs

Therefore we can conclude that the less efficient firm behaves more cautiously in the debt market whereas the more efficient firm behaves more aggressively.

Figures 4 and 5 show the output levels of firm i and firm j as a function of the marginal cost of the firm j. We can conclude the following:

Result 2. *The SPNE output of firm i, q_i^{**} is increasing with the rival's marginal cost, c_j. On the contrary, the SPNE output of firm j, q_j^{**}, is decreasing with the firm's marginal cost.*

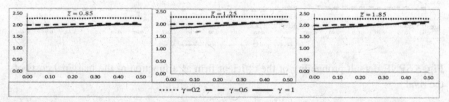

Fig. 3 SPNE output level of the more efficient firm as a function of the marginal costs of the rival

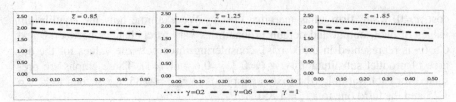

Fig. 4 SPNE output level of the inefficient firm as a function of its marginal costs

Therefore, as firm j becomes less efficient, the firm adopts a more conservative approach in the debt market and in the output market. The intuition for this result is that, an increase in the marginal production cost leads to a decrease in the marginal profit which implies a decrease in the debt and output levels. The more efficient firm has the opposite behavior, i.e. it becomes more aggressive in the debt market and in the output market. These effects are more pronounced for high levels of uncertainty, which increases the volatility of marginal profit.

Figures 6 and 7 show the default probabilities of firm i and firm j as a function of the marginal cost of firm j. These figures allow us to conclude the following:

Result 3. *The SPNE default probability of firm i, θ_i^{**} is increasing with the marginal cost of firm j. On the contrary, the SPNE default probability of firm j, θ_j^{**} is decreasing with the marginal cost of firm j. The change (decrease or increase) is more pronounced for high levels of the degree of product substitutability, γ.*

To understand the impact of changes in firm j marginal costs on its own default probability, one needs to consider both direct and indirect effects. The direct effect is positive. For given debt and quantity levels, an increase of the marginal cost of firm j increases its default probability. A first indirect effects results from the fact that the increase in the marginal cost of firm j leads to a more conservative behavior in the debt and output markets which implies a decrease in the default probability. A second indirect effect is related to the fact the efficient firm increases its quantity, which hurts the inefficient firm's profits and thus increases its default probability. Thus the total effect of increasing c_j on the default probability of firm j may be positive or negative, depending on which of the effects dominates. We conclude that it is the first indirect effect that dominates.

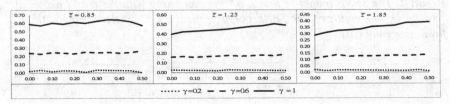

Fig. 5 SPNE default probability of the efficient firm as a function of the marginal costs of the rival

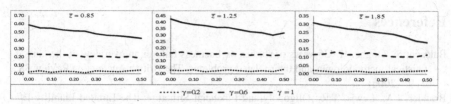

Fig. 6 SPNE default probability of the inefficient firm as a function of its marginal costs

5 Conclusion

The present work examined, analytically and numerically, how the market structure influence financial and product market decisions and, consequently, how it affects the default risk. We analyzed a two stage duopoly game model. In the first stage, firms decide the level of debt that maximizes the firm value and in the second stage of the game, firms decide on the optimal quantity that maximizes firm value for the shareholders. Due to the complexity of the problem, we had to solve the model analytically using GAUSS. We determined the SPNE of the debt levels, output levels and the default probabilities.

The result of the asymmetric duopoly reveal that the SPNE output decreases with the degree of product substitutability and with the firm's marginal cost of production and, on the contrary, it is increasing with the rival firm marginal cost. These results are similar to the ones obtained in traditional oligopoly models. Moreover, the equilibrium debt level of the less efficient firm is decreasing with its marginal cost while the most efficient firm has the opposite behavior. This is a quite interesting result as it tells us that the less efficient firm is more cautious and finances less with debt while the more efficient firm becomes «more aggressive» in the debt market. Another interesting result is that the default probability of the inefficient firm decreases as the firm becomes less efficient. This result is due to the existence of direct and indirect effects. On the one hand, for the same debt level, increasing the marginal cost of the firm is expected to lead to an increase on the default probability. On the other hand, since a decrease in efficiency leads to lower levels of debt and output, this leads to a decrease on the default probability. Overall, our results suggest that the default probability is greatly influenced by the financial and product market decisions of the firms, who optimally adjust their behavior to structural changes in the industry. Therefore a less favorable environment does not necessarily imply higher default probability, as the firm may respond by financing less with debt.

One possible extension of this work would be to incorporate the default costs and the debt tax savings effect in the model.

Acknowledgements. The authors thank the referees for their valuable suggestions, which have improved the paper. We are pleased to acknowledge financial support from Fundação para a Ciência e a Tecnologia and FEDER/COMPETE (grant UID/ECO/04007/2013).

References

Balcaen, S., Ooghe, H.: 35 years of studies on business failure: an overview of the classic statistical methodologies and their related problems. The British Accounting Review 38, 63–93 (2006)

Brander, J.A., Lewis, T.R.: Oligopoly and Financial Structure: The Limited Liability Effect. American Economic Review 76(5), 956–970 (1986)

Franck, B., Le Pape, N.: The commitment value of the debt: A reappraisal. International Journal of Industrial Organization 26(2), 607–615 (2008)

Haan, M.A., Toolsema, L.A.: The strategic use of debt reconsidered. International Journal of Industrial Organization 26(2), 616–624 (2008)

On the Optimization of Aircraft Maintenance Management

Duarte Dinis and Ana Paula Barbosa-Póvoa

CEG-IST, Instituto Superior Técnico, Universidade de Lisboa, Portugal

Abstract. Task scheduling and resource allocation problems have been the subject of intense research over the past decades, particularly within Operations Research. However, seldom optimization models have been proposed to address the aircraft maintenance management process in an integrated manner. Besides eliciting the problems of capacity planning, parts forecasting and inventory management, and task scheduling and resource allocation faced by aircraft MRO companies, this paper presents a short review on models that address each of the problems and discusses research opportunities within this field.

Keywords: aircraft maintenance, capacity planning, spare parts forecasting, inventory management, task scheduling, resource allocation.

1   Introduction

Maintenance is formally defined as the "combination of all technical, administrative and managerial actions during the life cycle of an item intended to retain it, or restore it to, a state in which it can perform the required function" (CEN 2001). The activity has evolved from a support activity that was seen as an inevitable part of production, to an essential strategic element needed for the accomplishment of business goals (Pintelon and Parodi-Herz 2008). Aircraft maintenance services provided by independent *Maintenance, Repair and Overhaul* (MRO) companies share many of the actions, policies, and concepts employed by traditional industrial maintenance, however some differences are to be mentioned. Most literature on maintenance address the subject from the perspective of the owner/operator of a complex system that needs maintenance (Mobley 2004), but to whom *maintenance* is not the core business (Marquez and Gupta 2006). This is not the case of independent aircraft MRO companies. Such companies *produce* maintenance services and in that sense their motivations and goals are different from those of the owner/operator of the aircraft (Kobbacy and Murthy 2008). The goal of the owner/operator is to retain or restore the reliability levels of an aircraft at a minimum cost, while the goal of an independent MRO is to achieve high service levels and to maximize profits. The motivations and goals of independent MROs are even different from those of airlines' subsidiary MROs,

© Springer International Publishing Switzerland 2015

A.P.F.D. Barbosa Póvoa and J.L. de Miranda (eds.), *Operations Research and Big Data,*
Studies in Big Data 15, DOI: 10.1007/978-3-319-24154-8_7

so are the difficulties faced. Traditionally, airlines' subsidiary MROs have been concerned with the availability of their own fleets, while independent MROs provide maintenance services to any operator in the market through outsourcing. Aircraft are for the former type of MROs what Fortuin and Martin (1999) categorize as "technical systems under client control". For these authors, there is usually a dedicated maintenance department providing maintenance services and managing spare parts inventories within the client organization in this case. As a result, a significant amount of information is available for the maintenance provider such as scheduled maintenance activities, times between failures, usage rates, and condition of the equipment. On the contrary, for independent MROs, aircraft resemble what Fortuin and Martin (1999) refer to as "end products being used by customers". In this latter case, much less information is available and such lack of information may result in occasional spikes of demand, both in manpower and spare parts.

The first source of information for the production planning of an aircraft MRO company must be the existence of reliable business forecasts regarding the number and type of products that are to be intervened in a given period of time. These forecasts allow the company to set the required manpower to meet the expected workload and to order the necessary materials in time. However, as aforementioned, the information that independent MROs possess on this matter is often scarce or unreliable. In addition to an estimate on the number and type of products expected to be intervened, a prediction regarding the scale of such interventions is also desirable. Any maintenance program contains periodic tasks that must be performed to keep the equipment in perfect working order. For aeronautical products, besides the replenishment of consumable materials and the replacement of spare parts and components that have reached their potential, the large majority of maintenance tasks include some kind of inspection. If in the course of these inspections an item is found damaged or below admissible tolerances, it must be repaired, or replaced altogether.

The set of tasks included in the maintenance program make up the *preventive maintenance* (also known as scheduled maintenance) (CEN 2001), while the repair or replacement tasks that result from inspections make up the *corrective maintenance* (also known as unscheduled maintenance) (CEN 2001). The duration of maintenance tasks is easily estimated since the maintenance programs from the *Original Equipment Manufacturers* (OEMs) set execution times for the tasks. Also the consumable materials, spare parts, and components are listed in the maintenance programs and can easily be purchased and stocked prior to the beginning of the intervention. However, this is not the case for the unexpected repair work. The required working time to fulfill the repair tasks and the need of parts that are found damaged or operating below functional limits cannot be easily predicted and is subject to a high degree of uncertainty. Materials that result from unexpected repair work often show a lumpy demand pattern (Ghobbar and Friend 2002), and their lead times can be higher than the immobilization periods of the aircraft themselves. Other factors that contribute to the increased complexity of the problem include the execution of certain tasks within confined spaces, according to the aircraft zones, and license restrictions to the employability of maintenance technicians. The possibility of adding extra resources to reduce the duration of tasks performed in confined spaces is limited, while each technician possesses only a certain number of licenses, thus being authorized to perform only a certain type of tasks, according to his/her technical skill, in a certain aeronautical product (Dijkstra et al. 1994).

The aircraft maintenance management problem is then quite complex, involving different aspects that can be grouped into three important sub-problems: (1) capacity planning of manpower; (2) spare parts forecasting and inventory management; and (3) task scheduling and resource allocation of scheduled and unscheduled maintenance tasks.

Relevant articles within Operations Research that address such sub-problems are hereby analyzed. The analysis does not intend to be an exhaustive review on all the models deriving from Operations Research applied to the aircraft maintenance problem. Instead, some examples of important papers addressing each sub-problem are presented.

2   Capacity Planning

Capacity planning, in this particular context, is the activity under which MRO companies set their manpower in a particular moment to face future demand (Budai et al. 2008).

Dijkstra et al. (1991) present their work on the development of a *Decision Support System* (DSS) for capacity planning at KLM, the Royal Dutch Airlines. The optimization problem is formulated as an integer program and solved using an approximation algorithm based on Lagrangian relaxation due to the NP-hardness of the problem. The DSS consists in a database module to store data about workload and workforce, an analysis module, which computes possible outcomes in the form of *scenarios*, and a graphical user interface. Particularly the analysis module, which uses optimization models, consists in routines to estimate workload, optimize the size and organization of the workforce, and evaluate the quality of the calculated workforce (Dijkstra et al. 1994). The objective function of the optimization routine is to determine the minimum number of maintenance engineers such that all jobs are performed, while the objective function of evaluation routine is to maximize the total number of jobs that can be performed given a certain workforce size and composition.

More recently, Yan et al. (2004) present an aircraft maintenance capacity planning model using flexible management strategies and considering multiple maintenance certificates. The authors employ three flexible strategies in order to increase the different degrees of freedom in the maintenance schedules: (1) flexible shifts, which allow the maintenance organization to set the optimal number of shifts and their starting times; (2) flexible teams (called *squads*), which allow the company to adjust the number of team members in response to change in demand; and (3) flexible working hours, which allow the company to set different number of working hours. The problem is formulated as a mixed integer program model with the objective of minimizing the total maintenance man-hours while satisfying the demand. The authors also develop a solution algorithm due to the size of the original problem. In Yang et al. (2003), the same authors formulated a similar model, however not considering multiple types of maintenance certificates.

Not specifically directed at aircraft maintenance contexts,.Ighravwe and Oke (2014) developed a non-linear integer programming model to minimize the number of maintenance personnel while maximizing productivity levels.

3   Spare Parts Forecasting and Inventory Management

According to Regattieri et al. (2005), there are two approaches to spare parts selection: (1) operational experience of an enterprise; and (2) application of forecasting techniques. Within the latter, the majority consist in statistical time series methods such as the *simple moving average* (SMA), *single exponential smoothing* (SES), and the *Croston's method* as witnessed by the works of Ghobbar and Friend (2002; 2003), Regattieri et al. (2005), and Romeijnders et al. (2012) in aircraft spare parts forecasting. Gutierrez et al. (2008) and Kourentzes (2013) use Neural Networks (NNs) as a forecasting technique for intermittent and lumpy demand profiles, but not specifically in aviation contexts.

To the authors' knowledge, optimization approaches have not yet been applied in conjunction with spare parts forecasting. They are however common in addressing inventory management problems. Atasoy et al. (2012) present a dynamic programming model to deal with the production/inventory problem under non-stationary stochastic supply uncertainty and availability of information about near future supply (called *advance supply information*). Wang (2012) uses a combined procedure of an enumeration and stochastic dynamic programming algorithm to optimize spare parts inventory based on a regular preventive maintenance interval. Zanjani and Nourelfath (2014) address the problem of spare parts logistics and operations planning for maintenance service providers. Firstly, the authors develop a deterministic mathematical programming model to find the optimal number of maintenance jobs that can be accomplished and the order quantity of spare parts that minimize procurement, inventory, and late delivery costs, taking into account the spare parts supply lead time. Secondly, the spare parts' demand uncertainty is modeled as a non-stationary stochastic process and the model reformulated as a multi-stage stochastic program with recourse. Gu et al. (2015) propose two non-linear programming models to find the optimal order time and order quantity while minimizing cost. In the aforementioned papers, with the exception of Atasoy et al. (2012), all mention the applicability of the methods to aircraft spare parts or are specifically motivated by this problem.

4   Task Scheduling and Resource Allocation

Task scheduling and resource allocation problems are addressed according to a multitude of techniques within Operations Research. Due to their relevance in the particular case of aircraft maintenance, and to their importance in the literature, job shop scheduling approaches are hereby presented. Other resource allocation approaches employed, or employable, within aircraft maintenance are also discussed.

Just as Ahire et al. (2000) mention, the problem of scheduling a set of preventive maintenance tasks with a given available workforce can be treated as a variation of the *Job Shop Scheduling Problem* (JSSP). The generic JSSP is set as follows: for a set of n jobs and a set of m machines (or resources), where each job consists in a sequence of operations and where each operation is performed continuously over a finite period of time in a given machine, the objective of the problem is to find a schedule that minimizes or maximizes certain performance metrics such as to minimize the total duration of the tasks, or maximize machine utilization (Yang et al. 2012). The problem has combinatorial characteristics and is considered NP-hard (Rajkumar et al. 2011; Karimi et al. 2012).

Solving approaches for the JSSP can be classified into two broad categories: (1) exact methods; and (2) approximation techniques. The exact method most commonly used for solving the JSSP has been the *Branch-and-Bound*, while *Genetic Algorithms* (GA), *Tabu Search* (TS), and *Simulated Annealing* (SA) are the most popular approximation techniques (Weckman et al. 2012).

Few articles were found implementing exact mathematical models for solving the JSSP within aircraft maintenance contexts. The work of Thörnblad et al. (2012) is one of the exceptions. The authors developed an exact mathematical model and applied it to a cell at Volvo Aero Corporation considering preventive maintenance activities. The cell consists of several work centers operating according to a *flexible job shop* logic. According to the authors, the limited research on exact mathematical models applied to JSSP is due to the high computational processing time that is required to solve some versions of the problem. However, the theoretical and practical advances made in mathematical optimization models and the evolution of computer hardware substantially decreased the computation time required, allowing the development of exact models for solving complex, and therefore realistic, instances of the JSSP.

Many of the articles on the JSSP present heuristics, meta-heuristics, and hybrid techniques for its resolution. Among them is the work of Ahire et al. (2000) who present a study on *Evolution Strategies* (ES) applied to preventive maintenance scheduling constrained by the available workforce. ES differ from GA in the sense that they are more concerned with phenotypes (behaviors, i.e. in the case of scheduling problems the schedule length or makespan), instead of genotypes (genes, i.e. the sequence of tasks) (Fogel 1995; Ahire et al. 2000). According to the authors, the optimal solution to the problem is a schedule in which all maintenance tasks are performed, the available workforce is efficiently used, and the minimum time to perform preventive maintenance is determined. For the experiments considered in the paper, ES performed more efficiently than exhaustive enumeration (EE) in finding optimal solutions with increasing problem size. Faster convergence to optimality was also obtained when compared to SA.

Meeran and Morshed (2012) present a hybrid GA with TS. The authors state that the main motivation for their work was the inadequacy of exact methods and heuristic methods to solve the JSSP *per se*, proposing an algorithm that incorporates the performance of a GA in the global search for a solution and TS in the local search.

Another possible approach to the described problem is to view it as a resource allocation problem. Compared to the JSSP, where emphasis is given on sequencing tasks for a certain number of resources, considering or not their capacity constraints, allocation problems reverse somehow the issue, focusing on the distribution of existing resources by the tasks to be performed.

Adida and Joshi (2009) developed a robust optimization approach for task scheduling and resource allocation under uncertainty. According to the authors, project rescheduling can lead to significant suboptimality due to increased costs. In order to overcome this issue, they develop a model using robust optimization techniques. They compared the performance of a deterministic model and the robust optimization model, claiming that the robust model does not generally need rescheduling for a certain level of uncertainty. The robust model also provides better protection against violations of finish time and project deadlines constraints.

Huang et al. (2010) employ a GA in the allocation of human resources on an aircraft maintenance problem. They assume a predefined sequence of maintenance tasks (called *items*) and then calculate an optimal allocation of resources to minimize the time span of the maintenance tasks.

The work of Safaei et al. (2011) address the problem of scheduling maintenance activities of a fleet of military aircraft. The authors formulate the model as a mixed-integer programming model (MIP), and solve it using a branch-and-bound method. The main objective of the model is to maximize fleet availability and the main constraint is the availability of skilled-workforce.

Bertsimas et al. (2014) propose a binary optimization framework to address the dynamic resource allocation problem. The authors claim that the approach has a number of benefits such as allowing modeling flexibility by incorporating different objective functions, being applicable to a wide variety of problems, and providing near optimal solutions fast for large-scale problems.

5   Conclusions and Future Research

This paper presents the aircraft maintenance problem from the perspective of independent MRO companies, and divides it into three main sub-problems: (1) capacity planning of manpower to face uncertain demand; (2) spare parts forecasting and inventory management; and (3) task scheduling and resource allocation of preventive and corrective maintenance tasks. A literature review is performed in optimization techniques that address each sub-problem.

The purpose of this paper is not to present an exhaustive review on applicable optimization approaches to the proposed sub-problems. Instead, examples of various techniques that have been applied to solve them individually are presented, from exact methods, to approximation algorithms.

However, no work was found addressing the three sub-problems in an integrated manner. To the authors' knowledge, capacity planning, spare parts forecasting and inventory management, and task scheduling and resource allocation problems have not been solved using a systematic and integrated methodology within Operations Research. By doing so, strategic, tactical, and operational decisions can be made coherently throughout the management process, improving the overall efficiency of the MRO companies through the reduction of spare parts inventories and overtime costs, and, ultimately, the on-time completion of the maintenance projects (Samaranayake and Kiridena 2012). This constitutes an extremely complex problem, of great importance at the industrial level, and that deserves further attention from the research community.

Acknowledgments. The authors acknowledge the Portuguese National Science Foundation (FCT) for the PhD grant PD/BD/52345/2013.

References

Adida, E., Joshi, P.: A robust optimisation approach to project scheduling and resource allocation. Int. J. Serv. Oper. Informatics 4, 169–193 (2009). doi:10.1504/IJSOI.2009.023421

Ahire, S., Greenwood, G., Gupta, A., Terwilliger, M.: Workforce-constrained Preventive Maintenance Scheduling Using Evolution Strategies. Decis. Sci. 31, 833–859 (2000). doi:10.1111/j.1540-5915.2000.tb00945.x

Atasoy, B., Güllü, R., Tan, T.: Optimal inventory policies with non-stationary supply disruptions and advance supply information. Decis. Support Syst. 53, 269–281 (2012). doi:10.1016/j.dss.2012.01.005

Bertsimas, D., Gupta, S., Lulli, G.: Dynamic resource allocation: A flexible and tractable modeling framework. Eur. J. Oper. Res. 236, 14–26 (2014). doi:10.1016/j.ejor.2013.10.063

Budai, G., Dekker, R., Nicolai, R.P.: Maintenance and production: A review of planning models. In: Kobbacy, K.A.H., Murthy, D.N.P. (eds.) Complex Syst. Maint. Handb., pp. 321–344. Springer, London (2008)

CEN EN 13306:2001 - Maintenance Terminology. European Standard. European Committee for Standardization, Brussels (2001)

Dijkstra, M.C., Kroon, L.G., Salomon, M., et al.: Planning the Size and Organization of KLM's Aircraft Maintenance Personnel. Interfaces (Providence) 24, 47–58 (1994). doi:10.1287/inte.24.6.47

Dijkstra, M.C., Kroon, L.G., van Nunen, J.A.E.E., Salomon, M.: A DSS for capacity planning of aircraft maintenance personnel. Int. J. Prod. Econ. 23, 69–78 (1991). doi:10.1016/0925-5273(91)90049-Y

Fogel, D.B.: Phenotypes, genotypes, and operators in evolutionary computation. In: Proc. 1995 IEEE Int. Conf. Evol. Comput. (1995). doi: 10.1109/ICEC.1995.489143

Fortuin, L., Martin, H.: Control of service parts. Int. J. Oper. Prod. Manag. 19, 950–971 (1999). doi:10.1108/01443579910280287

Ghobbar, A.A., Friend, C.H.: Sources of intermittent demand for aircraft spare parts within airline operations. J. Air. Transp. Manag. 8, 221–231 (2002). doi:10.1016/S0969-6997(01)00054-0

Ghobbar, A.A., Friend, C.H.: Evaluation of forecasting methods for intermittent parts demand in the field of: a predictive model. Comput. Oper. Res. 30, 2097–2114 (2003)

Gu, J., Zhang, G., Li, K.W.: Efficient aircraft spare parts inventory management under demand uncertainty. J. Air. Transp. Manag. 42, 101–109 (2015). doi:10.1016/j.jairtraman.2014.09.006

Gutierrez, R.S., Solis, A.O., Mukhopadhyay, S.: Lumpy demand forecasting using neural networks. Int. J. Prod. Econ. 111, 409–420 (2008). doi:10.1016/j.ijpe.2007.01.007

Huang, Z., Chang, W., Xiao, Y., Liu, R.: Optimizing human resources allocation on aircraft maintenance with predefined sequence. In: 2010 Int. Conf. Logist. Syst. Intell. Manag., ICLSIM 2010, vol. 2, pp. 1018–1022 (2010). doi:10.1109/ICLSIM.2010.5461109

Ighravwe, D.E., Oke, S.A.: A non-zero integer non-linear programming model for maintenance workforce sizing. Int. J. Prod. Econ. 150, 204–214 (2014). doi:10.1016/j.ijpe.2014.01.004

Karimi, H., Rahmati, S.H.A., Zandieh, M.: An efficient knowledge-based algorithm for the flexible job shop scheduling problem. Knowledge-Based Syst. 36, 236–244 (2012). doi:10.1016/j.knosys.2012.04.001

Kobbacy, K.A.H., Murthy, D.N.P.: An overview. In: Kobbacy, K.A.H., Murthy, D.N.P. (eds.) Complex Syst. Maint. Handb., pp. 3–18. Springer, London (2008)

Kourentzes, N.: Intermittent demand forecasts with neural networks. Int. J. Prod. Econ. 143, 198–206 (2013). doi:10.1016/j.ijpe.2013.01.009

Marquez, A.C., Gupta, J.N.D.: Contemporary maintenance management: process, framework and supporting pillars. Omega 34, 313–326 (2006). doi:10.1016/j.omega.2004.11.003

Meeran, S., Morshed, M.S.: A hybrid genetic tabu search algorithm for solving job shop scheduling problems: a case study. J. Intell. Manuf. 23, 1063–1078 (2012). doi:10.1007/s10845-011-0520-x

Mobley, R.K.: Maintenance Fundamentals, 2nd edn. Butterworth-Heinemann (2004)

Pintelon, L., Parodi-Herz, A.: Maintenance: An evolutionary perspective. In: Kobbacy, K.A.H., Murthy, D.N.P. (eds.) Complex Syst. Maint. Handb., pp. 21–48. Springer, London (2008)

Rajkumar, M., Asokan, P., Anilkumar, N., Page, T.: A GRASP algorithm for flexible job-shop scheduling problem with limited resource constraints. Int. J. Prod. Res. 49, 2409–2423 (2011). doi:10.1080/00207541003709544

Regattieri, A., Gamberi, M., Gamberini, R., Manzini, R.: Managing lumpy demand for aircraft spare parts. J. Air Transp. Manag. 11, 426–431 (2005). doi:10.1016/j.jairtraman.2005.06.003

Romeijnders, W., Teunter, R., Van Jaarsveld, W.: A two-step method for forecasting spare parts demand using information on component repairs. Eur. J. Oper. Res. 220, 386–393 (2012). doi:10.1016/j.ejor.2012.01.019

Safaei, N., Banjevic, D., Jardine, A.K.S.: Workforce-constrained maintenance scheduling for military aircraft fleet: a case study. Ann. Oper. Res. 186, 295–316 (2011). doi:10.1007/s10479-011-0885-4

Samaranayake, P., Kiridena, S.: Aircraft maintenance planning and scheduling: an integrated framework. J. Qual. Maint. Eng. 18, 432–453 (2012). doi:10.1108/13552511211281598

Thörnblad, K., Almgren, T., Patriksson, M., Strömberg, A.-B.: Mathematical optimization of a flexible job shop problem including preventive maintenance and availability of fixtures. In: Proc. 4th World P&OM Conf. / 19th Int. Annu. EurOMA Conf., Amsterdam, Netherlands, pp. 1–10, July 2012

Wang, W.: A stochastic model for joint spare parts inventory and planned maintenance optimisation. Eur. J. Oper. Res. 216, 127–139 (2012). doi:10.1016/j.ejor.2011.07.031

Weckman, G., Bondal, A.A., Rinder, M.M., Young, W.A.: Applying a hybrid artificial immune systems to the job shop scheduling problem. Neural Comput. Appl. 21, 1465–1475 (2012). doi:10.1007/s00521-012-0852-2

Yan, S., Yang, T.H., Chen, H.H.: Airline short-term maintenance manpower supply planning. Transp. Res. Part A Policy Pract. 38, 615–642 (2004). doi:10.1016/j.tra.2004.03.005

Yang, H., Sun, Q., Saygin, C., Sun, S.: Job shop scheduling based on earliness and tardiness penalties with due dates and deadlines: an enhanced genetic algorithm. Int. J. Adv. Manuf. Technol. pp. 657–666 (2012). doi: 10.1007/s00170-011-3746-z

Yang, T.H., Yan, S., Chen, H.H.: An airline maintenance manpower planning model with flexible strategies. J. Air Transp. Manag. 9, 233–239 (2003). doi:10.1016/S0969-6997(03)00013-9

Zanjani, M.K., Nourelfath, M.: Integrated spare parts logistics and operations planning for maintenance service providers. Int. J. Prod. Econ. 158, 44–53 (2014). doi:10.1016/j.ijpe.2014.07.012

Petroleum Supply Chain Network Design and Tactical Planning with Demand Uncertainty

Leão José Fernandes[1,2], Susana Relvas[2,*], and Ana Paula Barbosa-Póvoa[2]

[1] CLC Companhia Logística de Combustíveis, EN 366, Km 18,
2050-125 Aveiras de Cima, Portugal
[2] CEG-IST, Instituto Superior Técnico, Universidade de Lisboa, 1049-001 Lisboa, Portugal
susana.relvas@tecnico.ulisboa.pt

Abstract. The energy sector has dominated the world attention as recently arrived shale gas and slumping crude oil prices globally affect Petroleum Supply Chains (PSC). Fluctuating demand and prices are forcing companies to rethink existing distribution networks and production plans in order to improve network flexibility and cost efficiency. The current paper addresses this problem and presents a stochastic mixed integer linear program (MILP) for PSC design and planning under demand uncertainty that maximizes the expected net present value (ENPV) of a multi-entity multi-product PSC network. The decisions include locations, capacities, transport modes, routes, product affectations, inventories and fair-price costs and tariffs with economies of scale. The stochastic MILP designs a grassroots network with demand uncertainty for the Portuguese PSC, drawing important insights and research directions.

Keywords: Petroleum supply chain, Network design, Tactical planning, Stochastic MILP, Demand uncertainty, Expected value.

1 Introduction

The PSCs attract much attention as they shape the living standards and prices of nearly all goods (World Economic Forum [1]). In recent years, the fabric of this industry has changed dramatically. As crude oil prices doubled since 2009, many global companies invested in the pre-salt crude oil production where higher production costs were sustained by the increased prices. In recent years, however, the technology improvement accompanied by cost reduction in hydraulic fracking has made shale oil development economically viable. Investors hence moved into shale oil production in Northern America and many other countries. However, the fall in oil prices in 2014, which sharply reverted to the 2009 prices, has changed the economic equation and consequently the viability of these investments. Even shale oil whose production costs lowered significantly following the introduction

---

* Corresponding author.

© Springer International Publishing Switzerland 2015   59
A.P.F.D. Barbosa Póvoa and J.L. de Miranda (eds.), *Operations Research and Big Data*,
Studies in Big Data 15, DOI: 10.1007/978-3-319-24154-8_8

of new technology is being hit by the collapse of oil prices. This has created price, cost and demand volatility in the PSC, where companies and scholars try to improve the upstream and downstream efficiencies as seen in recent developments.

To deal with uncertainty, stochastic programing has been extensively used in transportation, logistics, finance and energy. Birge and Louveaux [2] presented two-stage recourse, multi-stage stochastic and stochastic integer programming methods besides Monte Carlo, approximating expectations and multistage approximation sampling methods for scenario generation. Some solution evaluation measures were proposed such as expected value of perfect information (EVPI), value of stochastic solution (VSS), expected value of the wait-and see problem (WS), expected recourse problem solution value (RP) and expectation of the expected value problem (EEV). Conejo et al. [3] present stochastic programming models for price, cost, supply, demand and equipment uncertainties in the electrical markets. Guillén et al. [4] propose a two-stage stochastic multi-objective MILP to maximize ENPV, demand satisfaction and minimize financial risk. Pareto stochastic curves are obtained for the SC tradeoff design configurations. Cardoso et al. [5] propose a stochastic MILP that maximizes ENPV for strategic planning with integration of reverse logistics. These models determine the installation, sizing and location of plants, warehouses and processes besides inventory levels and forward flows to retailers. The MILP in Cardoso et al. [5] additionally determines the reverse flows to retailers.

The oil supply, transformation and distribution under uncertainty was first studied in Escudero et al. [6] proposing a 2-stage stochastic scenario analysis and a deterministic equivalent model. Dempster et al. [7] developed deterministic and stochastic models for PSC logistics tactical and operational planning. These works proposed decomposition based on Augmented Lagrangian, Benders and Cholesky factorization for future research due to the very large scale of the problem. Later on, the optimal design of integrated PSC and biofuel operation under demand and price uncertainty was modeled in Tong et al. [8] with a two-stage stochastic MILP. The aim was to minimize expected overall costs while determining the installation and sizes of technology, transport affectations, and feedstock and product flow rates. Yue and You [9] proposed a bilevel MINLP for optimal design and planning of non-cooperative bio-fuel SC using Stackelberg game and generalized Nash equilibrium assumption. The bilevel MINLP is transformed into a single-level nonconvex MINLP using Karush–Kuhn–Tucker (KKT) conditions and globally optimized using a branch-and-refine algorithm. Facility location, sizing, technology selection, material flows and piecewise linearized prices are obtained for profit maximization of individual manufacturers.

Downstream PSC investment planning under demand uncertainty is modeled with a two-stage stochastic program in Oliveira and Hamacher [10]. A portfolio and timings of projects is selected that minimize costs while considering transportation navigability, operational capacity of rotating tanks, and demurrage costs. The sample average approximation (SAA) is used to generate scenarios. Later, Fiorencio et al. [11] develop a decision support system (DSS) to compare investments using a deterministic MILP that maximizes the NPV of a PSC while determining an investment portfolio of vessel and transportation capacity expansions along with their installation times.

Recently, Fernandes et al. [12] presented a multi-entity multi-product deterministic MILP for collaborative design and tactical planning of downstream PSC that maximizes the total network profits while determining the installation and operation of depot locations, resource capacities, transport modes and routes besides affectation of multi-period inventories per PSC stage and product flows on route to and between depots and clients. Fair-price costs and tariffs are determined based on economies of scale. Product independent equipment are modeled as continuous flow resources (ex. Pipeline, pumps, valves) while product dependent equipment are modeled as static flow resources (ex. vessels) where their flow rate is approximated in terms of rotations per period. Test results of the MILP based on the Portuguese PSC real data, identified the need for incorporating measures for handling the price, cost and demand uncertainties.

The present paper extends the work in Fernandes et al. [12] to a stochastic MILP for strategic and tactical planning of downstream PSCs under demand uncertainty. The stochastic MILP incorporates demand scenarios and maximizes ENPV for the PSC network while permitting storage depots to directly import product. Costs and tariffs are determined per entity, depot, route, product and period. The remainder sections present the problem description, the MILP model decisions, the stochastic mathematical formulation, the computational results based on the Portuguese PSC and finally conclusions and research proposals.

2 Problem Description

The downstream PSC as presented in fig. 1 is a network of refineries, distribution depots, transportation modes and routes that involves petroleum entities which purchase crude oil, produce or import petroleum products and finally export or commercialize and distribute product to retail customers aggregated in districts. In fig.1, the variables in grey are strategic decisions and the variables in black are tactical decisions. These are presented in detail in section 3.

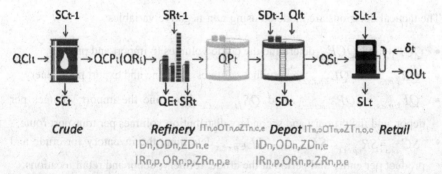

Fig. 1 The Downstream Petroleum Supply Chain

The petroleum entities design, construct and manage PSC networks with uncertainty, maximizing profits for the entire PSC wherein are considered various stages and self and partially owned installations. In this context, collaborative networks are the new paradigm where entities must balance participations and install and operate optimal depot and transport route locations, storage and transport capacities, determine efficient product transfer volumes, service tariffs and cost synergies, thereby improving the network efficiency and making the PSC member participation cost-effective for all entities. Along this paper, the design and planning grassroots problem is addressed where the storage depots, transportation infrastructures and entity participations are developed from scratch.

3   Model Decisions

Within the problem in study, strategic and tactical decisions are considered simultaneously. The first ones are related to the network design while the second ones relate to the planning of the network. The indices include refineries $i \in I$, depots $j \in J$, customers $k \in K$, products $p \in P$, entities $e \in E$, time period $t \in T$, scenarios $s \in S$, resources $r \in R$, transport modes $m \in M$, transport mode families $fm \in FM$, product families $fp \in FP$, starting nodes $h \in H = I \cup J$, destination points $l \in L = J \cup K$.

The problem's strategic decisions are modeled using binary and integer variables:

- ID_j and OD_j determine the depots to install and operate; $IT_{h,l,m}$ and $OT_{h,l,m}$ the transport routes to install and operate; $ZD_{j,e,u}$ and $ZT_{h,l,m,e,u}$ the operating thresholds per entity in depots and transport route to determine the piecewise linearized costs and tariffs per entity. (Binary variables)
- $IR_{j,fmr,fp}$ and $OR_{j,fmr,fp}$ determine the depot resource capacities to install and operate per transport mode and product; $ZR_{j,fmr,fp,e}$ the resource partitions to operate per depot, transport mode, product and entity in order to determine the piecewise linearized costs and tariffs per entity. (Integer variables)

The tactical decisions are modeled using non-negative variables:

- $QCI_{i,t,s}$ and $QCP_{i,t,s}$ determine the crude volumes to import and process;
- $QR_{i,p,t,s}$ and $QE_{i,p,t,s}$ the product volumes to refine and export per refinery;
- $QI_{j,p,e,t,s}$, $QP_{h,j,m,p,e,t,s}$, and $QS_{j,k,m,p,e,t,s}$ determine the import volumes per depot, and the primary and secondary distribution volumes per transport route;
- $SC_{i,t,s}$, $SR_{i,p,e,t,s}$, $SD_{j,p,e,t,s}$, $SL_{k,p,e,t,s}$ determine the inventory for crude and product per entity and period at the crude, refinery, depot and retail locations;
- $QU_{k,p,e,t,s}$ determine the unsatisfied demand.

4 Mathematical Formulation

The stochastic MILP uses the compact node-variable mathematical formulation which performs better than the scenario-variable formulation as identified in Conejo et al. [3]. The model maximizes the Expected Net Present Value (ENPV) in eq. (1) while considering uncertain demand. The ENPV is the sum of the product of each scenario probability ψ_s and respective NPV_s. Eq. (2) provides the NPV_s as the summation of the discounted profits obtained in each time period, using irr which is an arbitrary interest rate.

$$Max(ENPV) = \sum_s \psi_s \times NPV_s \tag{1}$$

$$NPV_s = \sum_t \mathrm{Pr}\,ofit_{t,s} \big/ (1+irr)^{t-1} \tag{2}$$

$$\mathrm{Pr}\,ofit_{t,s} = \sum_{e \in E} \left(\begin{array}{l} \sum_{i \in I} MR_{i,e,t,s} - \sum_{i \in I}\sum_{p \in p} CE_{i,p,e,t,s} - \sum_{j \in J}\sum_{p \in p} CI_{j,p,e,t,s} \\[2mm] - \sum_{p \in P}\left(\sum_{i \in I} CNC_{i,p,e,t,s} + \sum_{i \in I} CNR_{i,p,e,t,s} + \sum_{j \in J} CND_{j,p,e,t,s} + \sum_{k \in K} CNL_{k,p,e,t,s} \right) \\[2mm] + \sum_{k \in K}\sum_{p \in P} ML_{k,p,e,t,s} - \sum_{h \in H} TDE_{h,e,t,s} - \sum_{(h,l,m) | \exists \mu_{h,l,m}} TTE_{h,l,m,e,t,s} - \sum_{k \in K}\sum_{p \in P} CS_{k,p,e,t,s} \\[2mm] + \sum_{h \in H} MDE_{h,e,t,s} + \sum_{(h,l,m) | \exists \mu_{h,l,m}} MTE_{h,l,m,e,t,s} \end{array} \right) \tag{3}$$

Eq. (3) provides the multi-echelon PSC $\mathrm{Pr}\,ofit_{t,s}$ which aggregate: i) the refinery profit $MR_{i,e,t,s}$ less the export $CE_{i,p,e,t,s}$ and import $CI_{j,p,e,t,s}$ costs; ii) minus the inventory costs at the crude $CNC_{i,p,e,t,s}$, refinery $CNR_{i,p,e,t,s}$, depot $CND_{j,p,e,t,s}$ and retail $CNL_{k,p,e,t,s}$ stages; iii) plus the retail revenue $ML_{k,p,e,t,s}$ less the depot $TDE_{h,e,t,s}$ and transportation $TTE_{h,l,m,e,t,s}$ tariffs and shortage $CS_{k,p,e,t,s}$ costs; iv) plus the participation in depot $MDE_{h,e,t,s}$ and transportation $MTE_{h,l,m,e,t,s}$ profits the later calculated only for possible transport links $(h,l,m) | \exists \mu_{h,l,m}$.

The above costs, tariffs, revenues and profits are determined individually per entity e, per product p, per PSC stage location (refinery i, depot storage j, transportation m and retail k), per time period t and per scenario s for the shared network. The stochastic design and planning MILP considers market product prices, fixed refining and retail markups, detailed benchmarked costs and piecewise linearization to determine individual activity fair-price tariffs, in order to reduce costs and increase entity profits.

The PSC system is governed by scenario dependent crude and product volume balance equations at the refinery, storage depot, primary and secondary distribution networks and the retail stages. Multi-stage inventories are restricted by minimum and maximum constraints. The network volumes enforce the installation and operation of the required resource capacities per product and entity which may only exist in installed and operating storage depots and network infrastructures. A detailed mathematical formulation of the deterministic MILP for strategic and tactical planning of the downstream PSC may be consulted in Fernandes et al. [12].

5   Real Case Computational Results

The stochastic MILP is implemented using GAMS 24.4.1 64-bit and solved with MILP commercial solver CPLEX 12.6 on an Intel Xeon 2 GHz CPU with 16GB RAM and 16 parallel threads. The stopping criterion is 0.0% optimality gap or 24 hours of CPU time. The MILP is tested for a real example based on the Portuguese PSC data which was presented in Fernandes et al. [12]. Here, the demands, costs and prices are averaged into three five-year periods. The MILP models a grass-roots network for the Portuguese PSC with demand uncertainty. Besides the strategic and tactical decisions, the stochastic solution values are evaluated.

As shown in Fig. 2, a scenario tree is constructed for varying demand having a single scenario for the first period; three second period scenarios S1-S3 having 0%, 5% and 10% demand growth with 0.25, 0.5 and 0.25 probability; and nine third period scenarios S11-S33 with cumulative growth 0%, 5% and 10% and conditional probability 0.25, 0.5 and 0.25 based on the second period scenarios. Fig. 2 presents the NPV results for each of the demand scenarios.

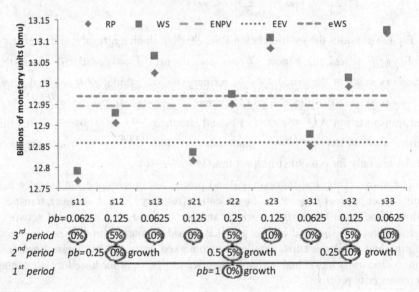

Fig. 2 NPV results for varying demand scenarios

Using the methodologies presented in Birge and Louveaux [1], the wait and see (WS), the recourse problem (RP) and the evaluation of the expected value (EEV) solutions are compared. The WS solution is the expected value obtained after solving each scenario using the deterministic model. The RP solution is the expected value obtained after solving the recourse problem. The EEV solution is the expected value obtained after fixing the first-stage variables to the deterministic solution for the average scenario S22. The NPV is seen to vary from 1.276,8 up to 1.311,7 bmu for the recourse solution and between 1.279 up to 1.312,4 bmu for the wait and see solution. The ENPVs obtained for the EEV, RP and WS problems are respectively 1.286, 1.295 and 1.297 bmu. On the network design Aveiro, Matosinhos, Azambuja, Sines and Barreiro are the depots selected in the EEV and RP solutions whereas Lisboa is additionally selected in the WS solution.

Fig. 3 compares the ENPV obtained for the WS, RP and EEV solutions. Here, the EVPI is the difference between the RP and EEV solutions which is the value a decision maker is willing to pay for perfect information. The VSS is the difference between the WS and RP solutions which measures the value of the stochastic solution over the deterministic equivalent. Based on the data, there is little improvement with the stochastic solution. This is justified as the major contributor to the ENPV is the refinery margin which is the least affected by domestic demand variations. The equidistance of the scenarios to the average scenario S22 is yet another reason. This suggests a detailed study of the demand history, to construct more realistic market scenarios so as to improve the quality of the stochastic solutions.

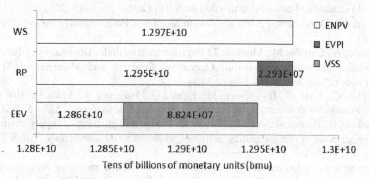

Fig. 3 Quality of the stochastic solution

Table 1 presents the model statistics for the WS, RP and EEV problem runs. The MILP is executed nine times for each scenario to obtain the WS solutions and is solved in 70280 seconds, the RP is solved in 86408 seconds and the EEV is solved in 29340 seconds. With only 3 periods and 9 scenarios, the problem is quite large as seen by the number of equations and variables, hence robust optimizing methods are seen to be more suitable to such a problem.

Table 1 Model Statistics

Problem	Equations	Total Variables	Binary Variables	Gap (%)	CPU(s)
WS x 9	30149	101841	4259	0.00	1005 – 27799
RP	134785	415685	4259	0.67	86408
EEV	88610	368053	3971	0.00	27799 + 1541

6   Conclusions

A stochastic MILP is presented for the optimal design and tactical planning of downstream PSC networks for ENPV maximization with uncertain demand. The stochastic MILP is an improvement to our earlier deterministic model. The MILP is tested for the Portuguese PSC real case, presenting the MILP solutions and performance results. The stochastic RP solution is compared to the EEV and WS solutions and determined the EVPI and VSS measures. These identify a potential to capture market variations and develop realistic scenarios in order to obtain better solutions. Consequently, the study of demand and price uncertainty, risk minimization and robust optimization for the PSC are proposed for future research.

Acknowledgments. The authors thank Fundação para Ciência e Tecnologia (FCT) and Companhia Logística de Combustíveis (CLC) for supporting this research.

References

1. World Economic Forum. Energy for Economic Growth - Energy Vision Update 2012. World Economic Forum in partnership with IHS CERA, pp. 1–45 (2012)
2. Birge, J.R., Louveaux, F.: Introduction to Stochastic Programming, 2nd edn. Springer (2010)
3. Conejo, A., Carrión, M., Morales, J.: Decision making under Uncertainty in Electricity Markets. International series in Operations Research and Management Science. Springer (2010). ISBN 978-1-4419-7420-4
4. Guillén, G., Mele, F., Bagajewicz, M., Espuña, A., Puigjaner, L.: Multiobjective supply chain design under uncertainty. Chemical Engineering Science 60(6), 1535–1553 (2005)
5. Cardoso, S., Barbosa-Póvoa, A., Relvas, S.: Design and planning of supply chains with integration of reverse logistics activities under demand uncertainty. EJOR 226(3), 436–451 (2013)
6. Escudero, L.F., Quintana, F.J., Salmerón, J.: CORO, a modeling and algorithmic framework for oil supply, transformation and distribution optimization under uncertainty. European Journal of Operational Research 114, 638–656 (1999)
7. Dempster, M.A.H., Pedron, N.H., Medova, E.A., Scott, J.E., Sembos, A.: Planning logistics operations in the oil industry. Journal of Oper. Research Society 51, 1271–1288 (2000)
8. Tong, T., Gong, J., Yue, D., You, F.: Stochastic Programming Approach to Optimal Design and Operations of Integrated Hydrocarbon Biofuel and Petroleum Supply Chains. ACS Sustainable Chemical Engineering 2, 49–61 (2013)
9. Yue, D., You, F.: Game-theoretic modeling and optimization of multi-echelon supply chain design and operation under Stackelberg game and market equilibrium. Computers & Chemical Engineering 71, 347–361 (2014)
10. Oliveira, F., Hamacher, S.: Optimization of the Petroleum Product Supply Chain under Uncertainty: A Case Study in Northern Brazil. Ind. Eng. Chem. Res. 51, 4279–4287 (2012)
11. Fiorencio, L., Oliveira, F., Nunes, P., Hamacher, S.: Investment planning in the petroleum downstream infrastructure. Intl. Trans. in Op. Res., 1–24 (2014). doi:10.1111/itor.12113
12. Fernandes, L.J., Relvas, S., Barbosa-Póvoa, A.P.: Collaborative Design and Tactical Planning of Downstream Petroleum Supply Chains. Industrial and Engineering Chemistry Research 53(44), 17155–17181 (2014)

Modeling Lot Sizing and Scheduling in Practice

Luis Guimarães, Gonçalo Figueira,
Pedro Amorim, and Bernardo Almada-Lobo

INESC TEC and Faculdade de Engenharia, Universidade do Porto, Portugal
{guimaraes.luis,goncalo.figueira,
amorim.pedro,almada.lobo}@fe.up.pt

Abstract. Lot sizing and scheduling by mixed integer programming has been a hot research topic in the last 20 years. Researchers have been trying to develop stronger formulations, as well as to incorporate real-world requirements from different applications. In this paper we illustrate some of these requirements and show how models have been adapted and extended. Motivation comes from different industries, especially from process and fast moving consumer goods industries.

1 Introduction

Integrating lot sizing and scheduling (L&S) is crucial for many companies. Lot sizing determines the timing and level of production to satisfy deterministic product demand over a finite planning horizon. Sequencing establishes the order in which lots are executed within a time period, accounting for the sequence dependent setup times and costs. The motivation for the integration of these two problems may come from: a) optimality point-of-view; the creation of more cost efficient production plans than those obtained when solving the two problems hierarchically by inducing the solution of the lot sizing problem in the scheduling level; feasibility point-of-view; in many production environments due to the extremely tight production capacity creating implementable production plans is challenging without this integrated approach.

The integration of L&S is required in production environments that usually share the following characteristics (Kallrath [2002]): multi-product equipment; significant sequence-dependent setup times and cleansing costs; divergent bill of materials; multi-stage production, with a known bottleneck; combined batch production and continuous operations. Furthermore, these industries often face strongly seasonal demand and capacity (usually constant) is insufficient to accommodate these variations. Under these conditions just-in-time systems cannot be implemented and a make-to-stock policy is economically preferable over investments in expanding capacity (Pochet [2001]).

A.P.F.D. Barbosa Póvoa and J.L. de Miranda (eds.), *Operations Research and Big Data*,
Studies in Big Data 15, DOI: 10.1007/978-3-319-24154-8_9

67

The recent developments in hardware and in the computational efficiency of modern commercial solvers have allowed researchers to propose more complex and realistic mathematical formulations for different L&S variants. Furthermore, the trendy and successful research line on mathematical programming-based heuristics benefit from stronger formulations. The contribution of this paper is to give an overview of the modeling features and extensions to address different requirements motivated by industrial applications of L&S. We cover extensions by analyzing: setup operations, synchronization of production resources, products features and planning process characteristics. We focus our attention on the modeling perspective rather than on solution approaches. Differently from previous reviews (see Drexl and Kimms [1997] and Jans and Degraeve [2008]) throughout the paper we provide examples from real world applications in which the reviewed extensions are needed describing the changes required to perform on a base model.

2   The Lot Sizing and Scheduling Model

The problem that considers the integration of sequencing decisions in the lot sizing and scheduling problem is known in the literature as the CLSD, Capacitated Lot Sizing Problem with sequence dependent setups, an extension of the original Capacitated Lot Sizing Problem (CLSP). The objective is to minimize the total expenditure in inventory and setups to fulfill demand over a finite planning horizon T, indexed by t. A plan simultaneously defines for every time period the production quantities and sequences for N products indexed by i and j. Demand is usually assumed to be known from forecasts and is to be met without backlog. Sequencing decisions are introduced since both the setup times and costs are dependent on the production sequence.

To formulate the basic form of the CLSD problem consider the following parameters.

d_{it} demand of product i in period t (units)

h_{it} holding cost of one unit of product i in period t

cap_t machine capacity in period t (time)

p_i processing time of product i

b_{it} upper bound on production quantity of product i in period t

st_{ij} time required (cost incurred) to perform a changeover from product i to
(sc_{ij}) product j

$\bar{s}_i$ start-up cost for product i

The decision variables to be optimized are: I_{it} the stock of product i at the end of period t, X_{it} the quantity of product i to be produced in period t, Z_{it}^b (Z_{it}^e) which is equal to 1 if machine is set up for product i at the beginning (end) of period t and T_{ijt} which takes the value of 1 if a changeover from product i to product j is performed in period t. The mixed integer mathematical formulation (MIP) for the basic CLSD reads:

$$\min \sum_{i,t} h_{it} \cdot I_{it} + \sum_{t,i,j} sc_{ij} \cdot T_{ijt} + \sum_{i,t} \bar{s}_i \cdot Z_{it}^b \tag{1}$$

$$\text{s.t.} \quad I_{i,t-1} + X_{it} = d_{it} + I_{it} \qquad\qquad\qquad \forall i,t, \tag{2}$$

$$\sum_i p_i \cdot X_{it} + \sum_{i,j} st_{ij} \cdot T_{ijt} \le cap_t \qquad\qquad \forall t, \tag{3}$$

$$X_{it} \le b_{it} \cdot \left(\sum_j T_{jit} + Z_{it}^b \right) \qquad\qquad \forall i,t, \tag{4}$$

$$\sum_i Z_{it}^b = 1 \qquad\qquad\qquad\qquad\qquad \forall t, \tag{5}$$

$$\sum_i Z_{it}^e = 1 \qquad\qquad\qquad\qquad\qquad \forall t, \tag{6}$$

$$Z_{it}^b + \sum_j T_{jit} = \sum_j T_{ijt} + Z_{it}^e \qquad\qquad \forall m,i,t, \tag{7}$$

$$\{(i,j) : T_{ijt} > 0\} \text{ does not include disconnected subtours} \quad \forall t. \tag{8}$$

$$X, I \ge 0, \quad Z \in \{0,1\}, \quad T_{jit} \in \{0,1\}. \tag{9}$$

The model is an adaptation of the model introduced in Guimarães et al. [2014]. The objective function (1) minimizes the sum of holding, setup and start-up costs. We assume that production costs are product and time independent, nevertheless such extension can be easily incorporated. The demand balancing constraints are described by (2). Production time together with the time lost in setup operations should not exceed the available capacity (3). Constraints (4) link production quantities to the machine setup state: production may only occur if a setup is performed in the period or the product is scheduled to be the first in the period. Constraints (5) and (6) ensure that the machine is set up for a single product in the beginning and at the end of each time period, while (7) keep trace of each machine configuration, balancing the flow of setups as follows. If there are no setups in period t the machine configuration at the end is the same as in the beginning of the period. If at least a setup is performed, for each product i three cases may appear: (i) more input than output setups, (ii) more output than input setups and (iii) equal number of input and output setups. In the first case the machine has to be set up for product i at the end of the period ($Z_{it}^e = 1$). The opposite scenario, the second case, forces the initial set up state for product i ($Z_{it}^b = 1$). The third case happens when the product is neither the first nor the last in the sequence, or it is not part of the production sequence of the machine in the period. Note that the current model assumes a complete setup state loss between two time periods, assigning a start-up cost to the first item produced in the period. Variables domain is defined in (9).

Finally, constraints (8) prevent disconnected subtours, i.e. sequences that start and end at the same setup state without this being the first. Guimarães et al. [2014] have shown in their study that the CLSD modeling efficiency is directly linked to the selection of the proper subtour elimination constraints. Without loss of generality we have modeled the single machine setting. However, such an extension is straight-forward by adding index m into variables X, Z^b, Z^e and T, summing machine's production quantities in constraints (2) and replicating the remaining constraints for all machines.

3 Modeling Setup Operations

Switching between production lots of two different products triggers operations, such as machine adjustments and cleansing procedures. These setup operations, which are dependent on the sequence, consume scarce production time and may cause additional costs due to, for example, losses in raw materials or intermediate products and equipment wear. However, correctly modeling these operations in strongly dependent on industrial features.

Setup Conservation between Time Periods. In some cases, the last machine setup state has to be carried over from one period to the next in order to properly account for the incurred setups. For instance, in process industries working on a 24x7 basis the borders in between periods simply appear due to modeling reasons, since no physical separation exists (note that the process is continuous). The basic CLSD model previously presented does not account for the so-called *setup carryover* and assumes a setup for every lot in the period. By introducing this extension we are not only improving solutions by decreasing the total setup cost but also doing an important contribution to the creation of feasible solutions in industries that face considerable setup times. To introduce this change, we impose $Z^e_{it} = Z^b_{i,t+1}, \forall i, t$ and $\bar{s}_i = 0, \forall i$ (Almada-Lobo et al. [2007]).

Period-Overlapping Setups. In the basic CLSD setups are performed entirely within a time period. In extremely tight capacity industries or whenever setups are substantially large when compared to the planning period length considering setup carryover may not be enough to provide feasible production plans. For instance, a color changeover in a furnace of a glass container manufacturer may take up to 3 days. In these scenarios allowing setups to overlap period's boundaries force them to be split between two or more time periods. In case two periods is enough, a setup is partially performed in period t and finished in period $t + 1$, which may delivery feasible valuable solutions in practice. This extension is called *setup crossover* and requires additional decision variables B_{ijt} indicating whether the cross over setup from period t to period $t + 1$ is from product i to j. (Menezes et al. [2011]).

Non-triangular Setups. In most production environments it is more efficient to change directly between two products than via a third product. This implies that in any optimal solution there is at most one production run for each product per time period. Under these conditions setups are said to obey to the triangle inequality. Nevertheless, in some cases contamination occurs when changing from one product to another forcing additional cleaning operations. If a 'cleansing' or shortcut product (often of lower grade) exists, contamination can be absorbed while producing such an item. Replacing the cleansing operations by a producing run often leads to the appearance of *non-triangular setups*. For example, in the animal nutrition industry contamination can occur in specific recipes (Clark et al. [2010]). In the presence of non-triangular setups, models have to allow for more than one production lot of each product per time period as it potentially reduces setup times and costs.

Despite being a simple concept, considering multiple production lots of the same product within each time period is a non-trivial extension. To start, one has to deal with integer T_{ijt} variables to determine the number of setups in period t from product i to j. In addition, connected subtours are now allowed in sequences which increases the complexity of subtour elimination constraints. Under these circumstances, a minimum lot size has to be imposed in order to eliminate unacceptable contamination or due to a modeling necessity to avoid fictitious setups to the short-cut product taking advantage of the setup time or cost reduction. As the production run may overlap more than one period (allowed by means of setup carryover), the calculation of the size of the lots must take into account contiguous periods.

Pre-defined Production Sequences. The motivation for this feature may come from modelling and industrial perspectives. Concerning the former, whenever the set of products to be produced becomes relatively large the model tends to be become intractable due to its size as the number of variables and subtour elimination constraints increases dramatically. With respect to the latter, in some industrial applications one need to fix a priori the sequence due to technological, costs or marketing reasons. A modeling alternative is to reformulate the production scheduling based on the selection of feasible production sequences (Haase and Kimms [2000]). Variables T_{ijt} are dropped from the model and a new set of decision variables is created to define the sequence selection. Additionally, constraints (5)-(8) are replaced by imposing the selection of a single sequence in each time period. Naturally, the creation of production sequences is an important ingredient in this procedure and can be approached using, for instance, a column generation algorithm (Guimarães et al. [2013]). This extension also allows modeling specific production sequences required in practice by constraining the possible set.

It must be defined the set of all S_t feasible production sequences to schedule products on the machine in period t, indexed by $s = 1, \ldots, S_t$. Binary decision variables W_{ts} capture whether sequence s is chosen in period t, whereas parameters e_{is} makes the link between the product and the sequence, defining whether the machine is ever set up for product i in sequence s.

Family Setups. Sometimes, there are important scale differences among the setups performed. In many practical situations, products are grouped into families implying that a changeover between products of the same family is much less costly than a changeover between products of different families. Changing between product families are often referred as *joint* or *major setups* and between items of the same family as *minor setups*. Another important difference can be associated with the setup structure, as sequence dependent setups may only apply when changing between families and sequence independent setups are present when changing between items of the same family. This scenario is closely related to the concept of recipe often found in fast moving consumer food goods. Changing between recipes requires major setups and between items of the same recipe minor setups. To model these production contexts we may apply an adaptation of the block planning formulation (Günther et al. [2006]) where each block is a recipe and inside each recipe the sequence of products is set *a priori*. To do so, the T_{ijt} variables have to be redefined and only applied to the changeovers between families, and new sequence independent setup variables need to be introduced for the product level.

4 Synchronization of Resources

Synchronizing resources may be critical at the operational level, where plans need to be more detailed. This is particularly important in the presence of shifting bottlenecks or when the schedule of a resource may impact the overall production costs. This can be the case of multi-stage production environments and also secondary scarce resources, some examples are given in Table 1.

Table 1 Production stages and secondary resources in different industrial environments

	Glass Container	Soft Drinks	Beer	Dairy	Pulp and Paper	Spinning	Foundry	Animal Nutrition
First Stage	Glass Furnaces	Tanks	Fermentation Tanks	Pasteurization Tanks	Digestor	Fiber Blending	Alloy Furnaces	Pre-mix machines
Second Stage	Molding lines	Filling Lines	Filling Lines	Packaging Lines	Paper Machine	Spinning Machines	Moulding lines	Mixer Machines
Secondary Resources	Moulds		Intermediate tanks		Pulp Tanks		Moulds	

When many intra-period events have to be captured, many constraints related to resources' synchronization may be difficult or even impossible to incorporate in the CLSD, since it allows for multiple products to be produced. Small-bucket models that partition the CLSD time periods into smaller ones and limit the production to one setup per time slot may thus be required/desirable. Although being more computationally expensive compared to large-bucket models like CLSD, they are more flexible when grasping resource states inside each time period. For more detail on the modeling properties of these models we refer to Belvaux and Wolsey [2001]. The general lot sizing and scheduling problem (GLSP) is the most flexible among small-bucket models [Fleischmann and Meyr, 1997].

Batch vs. Continuous Processes. When considering more than one production stage, it is important to take into account the way process material flows across production resources. Resources may operate continuously or in batches. In the former there is always material coming in and going out, and therefore there is no lead time between stages. This is the case of industries such as glass container and industrial metals. Batch processes require the material to stay inside the resource for some time. Only then it can start feeding subsequent stages. This occurs in industries such as beverages and dairy products, where the upstream process takes place in batches. The chemical and paper industries have both types of production, depending on the specific equipment installed at the plants. In continuous materials flows CLSD can be adapted to link the start and end of each production lot in each level such as for instance in the case of glass container industry. Small-bucket models are able to consider a more realistic flow between stages in batch production and when additional requirements emerge like in the case of pulp and paper (Figueira et al. [2013]).

Intermediate Buffers. Production stages can be connected directly or have intermediate buffers in between. If connected directly, we have to know exactly when the production of each item starts in the second stage, so that it can be synchronized with the material produced in the first stage (see Camargo et al. [2012]). When an intermediate buffer exists, the mass balances need to be evaluated whenever a changeover occurs (as different products will consume the input material at different rates), in order to ensure that the buffer limits are never violated. In both cases small-bucket models are more flexible than large-bucket models.

Scarce Resources. The synchronization of secondary resources can also be important. For instance, the timing of setups may be a relevant feature to model in certain production environments. In the presence of parallel unrelated machines the existence of a common setup operator may prevent two setups to overlap in time. This is the case when a specialized team is responsible for all the machine adjustments required when performing a changeover or a special tool is required to execute the setups. Moreover, equipments such as dies or molds may be required and shared by several products during their production runs. In many cases, the limited number of these equipments due to their high cost also forces their usage to be carefully planned. Under these conditions, ignoring the timing of setups can produce solutions impossible to implement in practice. To model explicitly the timing of setups new decision variables are introduced representing the start and end of the setups in the machines. Besides capturing the sequence of the production runs in the machines, the model has also to sequence the usage of the different scarce resources over time (Almeder and Almada-Lobo [2011]). Decision variables $W_{tmm'k}$ indicate whether the resource k is used on machine m after being used on machine m'. The starting and ending times of the attachment of a given resource to a machine in each period must be traced.

Variable Production Rates. Being able to change the production rates of certain resources at specific points in time may help coordinating different the stages and resources. This feature may require production rates to be modeled explicitly or just implicitly. In the latter the production quantity is constrained by the maximum and/or minimum rates. The former uses a variable to explicit model the production rate. This is needed for instance in the case where production rates are limited to a certain variation. Often these rates must also be as steady as possible and thus their change should be minimized in the objective function. Nevertheless, an explicit variable for production rates can make the model non-linear. In that case, the model should be linearized by using a discrete grid of rates or by leaving the rates smoothness to a post-optimization phase.

5 Products' Features

The properties of some products may also require changes in the model to fit reality.

Perishability. One of the most relevant features is perishability. Not accounting for product deterioration over time and/or shelf life can lead to significant spoilage costs. In practice it is required to trace the inventory age over the planning horizon. To capture this age-dependent inventory the simple plant location (SPL) reformulation can be used. The new production decision variables define simultaneously the period of production and of consumption allowing to capture the age of inventory. This extension not only allows to limit the use of stock during shelf life, but also to model delivery freshness and spoilage amounts, which are key features in the dairy industry (Amorim et al. [2013]).

Manufacturing Constraints. Products' manufacturing constraints can also impact the modeling decisions. These constraints may impose for certain products bounds on the minimum, maximum or fixed size of production runs (see Kang et al. [1999]) and can emerge from multiple reasons: quality assurance, materials consumptions, capacity of resources and utilization, and production already released. In the presence of setup carryover production runs can be split among two periods and the additional lot size constraints have to properly address this issue. Moreover, when considering a fixed production run size besides dropping the X_{it} variables from the formulation since the number of setups automatically define the production quantity, multiple production runs for the same product can also occur in the optimal solution. Thus, this extra requirement leads to formulations close to the ones considered for the case of non-triangular setups.

6 Planning Process

L&S is one of the most important and challenging key processes within the production planning of an industrial company, which is done under a hierarchical process with several echelons, each containing different aims and planning horizons. According to the Supply Chain Planning (SCP) matrix (e.g. Fleischmann and Meyr [2003]), L&S is of short/medium-term scope and is placed between Master Production Scheduling and Operational Scheduling. For a smooth implementation of the plans, simply solving those levels sequentially is not sufficient. The interaction demands the upper level to consider the characteristics of the lower level in its decision. On the other hand, lower level decisions are constrained by the instructions of the upper level.

Rolling Horizon. Much of the research in lot sizing and scheduling problems has concerned just the optimization of its static version. Better solutions are usually applied on a rolling horizon fashion. Moreover, in practice planners try to follow this scheme (even if not in a systematic way) to avoid the myopic nature of the static variants (specially in between two planning horizons). The basic idea is to create a plan for a specific planning horizon, but only the first period(s) is(are) executed (e.g. Clark and Clark [2000]). The remaining periods can be relaxed (e.g. not sequencing the production lots within each period) as they are re-optimized and updated as the horizon is rolled forward and new data gets available. A few key parameters influence the success of the rolling horizon approach, such as the planning horizon length, the planning frequency, the number of frozen periods and the way the decision variables are frozen (e.g. fixing setup-related variables and/or quantities). There is naturally a trade-off between flexibility, cost and stability/nervousness of the plans.

With regards to the key performance indicators (KPIs) that drive the planning process, the standard goal is to maximize met demand in the most cost-effective manner. The best trade-off between setup costs and holding costs under tight capacities is sought. Notwithstanding, other KPIs may emerge in different applications.

Demand Satisfaction Related Costs. Demand for individual products are aggregated per time period, and come from actual placed orders or forecasting systems. In case unmet demand is allowed, costs related to backlog or to lost sales are usually considered in deterministic lot sizing and scheduling. Note that in stochastic variants – not covered here – customer service related measures are explicitly incorporated. In some practical applications, companies can not backlog for more than one planning period, and the respective demand is converted into lost sales. In case of perishable goods, one needs to incorporate discarding costs. The spoilage costs – incurred whenever stock is held beyond the product's shelf-life – are usually defined as opportunity costs coming from the potential revenue yielded by the product to be discarded.

Setup Related Costs. Besides the changeovers from one individual product to another that cause sequence-dependent setup costs (and times), in L&S there are often other types of costs related to setups. The shutdown of a production resource (machine, furnace, ...) may incur in costs that are penalized. The cost of preserving a setup configuration – known as standby costs – without occurring production happens in some industries (e.g. beverage case). Setup costs may also appear not in product changeovers, but in scarce resources (secondary 'valuable' resources that need to be utilized during setup or production events).

Production Related Costs. The trade-off between setup and holding costs is sometimes extended to production costs. This is specially relevant in the presence of an heterogeneous multi-machine environment per production stage (e.g. machines of different technologies and processing rates). Additionally, in capital intensive industries the main operational driver is to maximize the facilities throughput (e.g.

glass tonnage produced in furnaces at the glass container industry and virgin pulp produced in digesters at the pulp and paper industry). In other cases, in order to promote the steadiness of the production resources, one needs to minimize the variation of the respective processing rates along the planning horizon.

Stock Related Costs. Stock holding costs, which may be time and product dependent, serve to penalize the production in advance to cope with scarce capacity. Stocks of intermediate or final products are differentiated according to the underlying bill-of-materials. In some industrial applications, as important as minimizing the stock carried at a certain point in time, is to guarantee the right balance of attributes of the intermediate products in inventory. For instance, in the spinning industry, the right combination of different fiber attributes is mandatory to achieve an homogeneous blend. Here, the variation of quality attributes between two consecutive fiber blends needs to be minimized.

7   Discussion

Lot sizing and scheduling has been a very active field of research and researchers have devoted an increasing attention in extending modeling approaches to incorporate additional realism. This process has allowed to accurately represent the planning problems faced by many industrial sectors. Nevertheless, there are still numerous areas for potential future research. The synchronization between stages needs further work especially in the presence of lead times. This can boost the use of these models in more complex production environments. The introduction of uncertainty is for sure another promising field of future research. Despite their operational/tactical nature, many of the model parameters are likely to suffer from variations when applying the plans in practice. Understanding the effects of such variability can help managers in their daily planning activities. Finally, the integration of lot sizing and scheduling with other supply chain related problems especially demand planning can further leverage the effective use of the tight capacity.

References

Almada-Lobo, B., Klabjan, D., Oliveira, J.F., Carravilla, M.A.: Single machine multi-product capacitated lot sizing with sequence-dependent setups. International Journal of Production Research 45(20), 4873–4894 (2007)

Almeder, C., Almada-Lobo, B.: Synchronisation of scarce resources for a parallel machine lotsizing problem. International Journal of Production Research 49(24), 7315–7335 (2011)

Amorim, P., Costa, A.M., Almada-Lobo, B.: Influence of consumer purchasing behaviour on the production planning of perishable food. OR Spectrum 36(3), 669–692 (2013)

Belvaux, G., Wolsey, L.A.: Modelling practical lot-sizing problems as mixed-integer programs. Management Science 47(7), 993–1007 (2001)

Camargo, V.C.B., Toledo, F.M.B., Almada-Lobo, B.: Three time-based scale formulations for the two-stage lot sizing and scheduling in process industries. Journal of the Operational Research Society 63(11), 1613–1630 (2012)

Clark, A., Morabito, R., Toso, E.: Production setup-sequencing and lot-sizing at an animal nutrition plant through atsp subtour elimination and patching. Journal of Scheduling 13(2), 111–121 (2010)

Clark, A.R., Clark, S.J.: Rolling-horizon lot-sizing when set-up times are sequence-dependent. International Journal of Production Research 38(10), 2287–2307 (2000)

Drexl, A., Kimms, A.: Lot sizing and scheduling - survey and extensions. European Journal of Operational Research 99(2), 221–235 (1997)

Figueira, G., Oliveira Santos, M., Almada-Lobo, B.: A hybrid VNS approach for the short-term production planning and scheduling: A case study in the pulp and paper industry. Computers & Operations Research 40(7), 1804–1818 (2013)

Fleischmann, B., Meyr, H.: The general lotsizing and scheduling problem. OR Spektrum 19(1), 11–21 (1997)

Fleischmann, B., Meyr, H.: Planning hierarchy, modeling and advanced planning systems. In: Graves, S., de Kok, A. (eds.) Supply Chain Management: Design, Coordination and Operation. Handbooks in Operations Research and Management Science, vol. 11, pp. 455–523. Elsevier (2003)

Guimarães, L., Klabjan, D., Almada-Lobo, B.: Pricing, relaxing and fixing under lot sizing and scheduling. European Journal of Operational Research 230(2), 399–411 (2013)

Guimarães, L., Klabjan, D., Almada-Lobo, B.: Modeling lotsizing and scheduling problems with sequence dependent setups. European Journal of Operational Research 239(3), 644–662 (2014)

Günther, H.-O., Grunow, M., Neuhaus, U.: Realizing block planning concepts in make-and-pack production using MILP modelling and SAP APO. International Journal of Production Research 44(18-19), 3711–3726 (2006)

Haase, K., Kimms, A.: Lot sizing and scheduling with sequence-dependent setup costs and times and efficient rescheduling opportunities. International Journal of Production Economics 66(2), 159–169 (2000)

Jans, R., Degraeve, Z.: Modeling industrial lot sizing problems: a review. International Journal of Production Research 46(6), 1619–1643 (2008)

Kallrath, J.: Planning and scheduling in the process industry. OR Spectrum 24(3), 219–250 (2002)

Kang, S., Malik, K., Thomas, L.J.: Lotsizing and scheduling on parallel machines with sequence-dependent setup costs. Management Science 45(2), 273–289 (1999)

Menezes, A., Clark, A., Almada-Lobo, B.: Capacitated lot-sizing and scheduling with sequence-dependent, period-overlapping and non-triangular setups. Journal of Scheduling 14(2), 209–219 (2011)

Pochet, Y.: Mathematical programming models and formulations for deterministic production planning problems. In: Jünger, M., Naddef, D. (eds.) Computat. Comb. Optimization. LNCS, vol. 2241, pp. 57–111. Springer, Heidelberg (2001)

Improving the Robustness of Bus Schedules Using an Optimization Model

Joana Hora, Teresa Galvão Dias, and Ana Camanho

Faculdade de Engenharia da Universidade do Porto, Portugal
CEGI – INESC TEC

Abstract. This study pursues the operational improvement of urban transportation services. Non-foreseen events lead to the occurrence of delays, which are further propagated during the daily operations of bus services. This paper applies an optimization model to obtain robust schedules of bus lines. The model builds a new schedule which minimizes delays and anticipations from a set of observations. The decision variables are the slack time to be allocated at each segment of two subsequent stops. The solutions obtained are assessed with two robustness measures: price of robustness (i.e. the deviations from schedule) and the percentage of absorbed delays. The results obtained in a real-world case study (a bus line operating in Porto) are promising.

Keywords: Robust Optimization, Bus Schedules, Urban Transports.

1   Introduction

The demand for urban transports, namely buses, has increased in the last decades together with the growth of cities (Ibarra-Rojas et al. 2015). This created the challenge of enhancing the performance of urban transportation systems, leading to improved accessibility in urban areas (Schmid 2014; Farahani et al. 2013). Citizens expect bus services to satisfy their needs efficiently, with high quality, reduced costs, reduced travel time, punctuality and high availability (Ceder and Wilson 1986; Zhao and Zeng 2008). The appropriate planning of bus systems is expected to ensure adequate service provision with low tariffs (Ceder 2007).

This study explores the optimization of bus schedules. Bus scheduling pursues the minimization of the operating costs, waiting time for passengers, and inconveniences caused by delays and anticipations from the schedule (Liu and Wirasinghe 2001). This must be achieved without disregarding the dynamics of the system, particularly the coordination of bus routes, the occurrence of unforeseen events causing instability (Hill 2003), specific customer needs, or the features of the geographic area related to demand and topology (Szeto and Wu 2011; van Oudheusden and Zhu 1995; Mulley and Ho 2013).

© Springer International Publishing Switzerland 2015
A.P.F.D. Barbosa Póvoa and J.L. de Miranda (eds.), *Operations Research and Big Data*,
Studies in Big Data 15, DOI: 10.1007/978-3-319-24154-8_10

Daily operations of bus systems are highly exposed to uncertainty arriving from non-foreseen events, such as traffic, weather conditions, mechanical failures, road block, strikes or unexpected peaks of demand (van Oudheusden and Zhu 1995; Farahani et al. 2013; Naumann et al. 2011). Uncertainty leads to the delay of trips, causing a severe deterioration on the quality of the service provided and an increase in operational costs (Naumann et al. 2011). *Primary Delays* occur due to disruptions that cannot be prevented (e.g., traffic or road block), causing the later arrival of a trip that has departed on time. A *Secondary Delay* occurs when an already delayed bus trip leads to a delayed start of the next bus trip. This inheritance of delays is called *Delay Propagation,* resulting from dependencies of consecutive trips whose slack times are not enough to absorb previous delays (Naumann et al. 2011).

Managing uncertainty within bus transportation systems can be addressed with robust planning (during the planning phase) or with dynamic re-planning (during the daily activity). This paper addresses the robust planning of bus schedules by optimizing the distribution of slack time in bus schedules.

Following a strict definition of the concept, a strictly robust schedule performs well in all scenarios that may occur under uncertainty, even in the worst case scenario (Kouvelis and Yu 1997). It diverges from schedules obtained via Stochastic Optimization, as the decision process does not consider uncertainty to follow a specific probabilistic distribution (Bertsimas et al. 2011; Kouvelis and Yu 1997).

Strictly robust solutions are considered too conservative, as they imply a great loss of optimality in order to guarantee robustness (Goerigk et al. 2011). Alternative approaches such as Light Robustness (Fischetti and Monaci 2009) return solutions complying with a certain qualification whilst maximizing robustness (Goerigk et al. 2011), considering an initial specification of "robustness goal" and "maximum objective function deterioration" to be accepted in the model (Fischetti and Monaci 2009). The majority of Robust Optimization applications in transportation studies intend to address the specific needs of each problem, leading to different robustness definitions and measurements.

This study applies an optimization model to obtain robust bus schedules. The model builds a new schedule which minimizes delays and anticipations from a set of observations. The set of real observations is used to define the range in which uncertainty fluctuates. This is a standard approach in classical robust optimization (Liebchen et al. 2009). The decision variables are the slack time to be allocated at each segment of two subsequent stops. The model was applied to a real-world case study. The results obtained are discussed and future research directions are proposed in the conclusions.

2 An Optimization Model for Robust Bus Scheduling

This section presents the formulation of an optimization model that aims to obtain a bus schedule that minimizes delays and anticipations observed in the Time Control Points (TCP) monitored for a given bus route. This involves a comparison between the real time that the bus actually arrived at the TCP for a set of observations collected during the monitoring period and the time stated in the original schedule. The optimization model shown in (1) searches for the values of slack time to allocate between two adjacent TCPs ($\tau_{m,m+1}$) in the schedule under construction. This model can be equivalently converted by standard techniques into a Mixed Integer Linear Programming (MILP) model (Garfinkel and Nemhauser 1972).

$$Min\ F(\tau, \gamma_1, \gamma_2) = \sum_{m=1}^{M} \sum_{k=1}^{K} \frac{\gamma_1 \cdot \max(0, SD_{m,k}) + \gamma_2 \cdot \max(-SD_{m,k}, 0)}{K} \tag{1}$$

$$SD_{1,k} = 0 \tag{1.1}$$

$$SD_{m+1,k} = E(T_{m,m+1}) + \tau_{m,m+1} - T_{m,m+1,k} + (1 - \beta_{m,m+1}) * SD_{m,k} \tag{1.2}$$

$$E(T_{m,m+1}) + \tau_{m,m+1} \geq N_{m,m+1} \tag{1.3}$$

$$\sum_{m=1}^{M-1}(E(T_{m,m+1}) + \tau_{m,m+1}) = \Psi \tag{1.4}$$

Where:
$M \in \mathbb{Z}$: number of Time Control Points (TCPs); m=1,…,M-1;
$K \in \mathbb{Z}$: number of real observations of the system;
$\gamma_1 \in \mathbb{R}$: weight for earlier arrivals, $\gamma_1 \in [0,1]$;
$\gamma_2 \in \mathbb{R}$: weigh for later arrivals, $\gamma_2 \in [0,1]$;
$\Psi \in \mathbb{R}$: total time for the daily trip (in time units).
$\tau = [\tau_{m,m+1}] \in \mathbb{R}^{M-1}$: vector storing the slack times to be allocated in each segment between two consecutive TCPs $[m, m + 1]$ (decision variables - in time units);
$\beta = [\beta_{m,m+1}] \in \mathbb{R}^{M-1}$: vector of dimension $(M - 1)$ storing the Adjustment Factor of Driver between each pair of adjacent TCP (input data, percentage);
$E(T) = [E(T_{m-1,m})] \in \mathbb{R}^{M-1}$: vector of dimension $(M - 1)$ storing the expected value of Travel Time in each segment (in time units);
$N = [N_{m,m+1}] \in \mathbb{R}^{M-1}$: vector storing the minimum travel time needed by a vehicle to travel between two adjacent TCP (input data, in time units);
$T = [T_{(m,m+1),k}] \in \mathbb{R}^{(M-1) \times K}$: array of dimension $(M - 1) \times K$ storing the Travel Time in each segment $[m, m + 1]$ at each real observation k (input data, in time units);
$SD = [SD_{m,k}] \in \mathbb{R}^{M \times K}$: array of dimension $M \times K$ containing in each position the Schedule Deviation at TCP m under observation k (in time units); $SD_{m,k} > 0$ indicates a delay, $SD_{m,k} < 0$ indicates an anticipation and $SD_{m,k} = 0$ indicates a perfect compliance with the schedule.

Model (1) is based in the formulation proposed by Yan et al. (2012). The objective function quantifies earlier and later deviations from the schedule, with associated weights γ_1 (earlier arrivals) and γ_2 (later arrivals). The component associated with the minimization of the absolute value of deviations, included in the formulation of Yan et al. (2012), was removed from our formulation as the quantification of deviation variability was considered to have a high degree of redundancy with the other two components of the objective function, and would increase considerably the computation complexity of the optimization model (changing from quadratic complexity to cubic complexity).

Constraint (1.1) assumes the system to start on time ($SD_{1,k} = 0$). Constraint (1.2) defines the Schedule Deviation (SD_{mk}), quantifying deviations with respect to the planned schedule, for each observation k in each TCP m. Each element $SD_{m+1,k}$ is defined as the difference between the schedule under construction ($E(T_{m,m+1}) + \tau_{m,m+1}$) and the actual arrival time ($T_{(m,m+1),k}$), plus recovery reached by drivers behavior ($(1 - \beta_{m,m+1}) * SD_{m,k}$). These constraints are similar to the formulation of Yan et al. (2012).

Constraint (1.3) ensures that the travel time between two adjacent TCPs in the schedule under construction ($E(T_{m,m+1}) + \tau_{m,m+1}$) is greater or equal to the minimum travel time between these TCPs ($N_{m,m+1}$) defined by the analyst. Constraint (1.4) establishes that the total duration of the daily trip in the new schedule ($\sum_{m=1}^{M-1}(E(T_{m,m+1}) + \tau_{m,m+1})$) is equal to the parameter Ψ. These two constrains are new features proposed in our formulation.

Our formulation considers the optimization of non-cyclic bus schedules adopting a daily perspective. This way the model accounts for fluctuations in demand and uncertainty occurring at different moments of the day. The assessment of the robustness of each schedule is based on two measures: Price of Robustness (PoR) and Percentage of Absorbed Delays ($PoAD$).

The PoR indicator, shown in (2), measures the reduction of deviations between the new and current schedules, in percentage. It is calculated as the ratio between the objective function value of the schedule under construction ($F(\tau, \gamma_1, \gamma_2)_R$) and the objective function value of the current schedule ($F(\tau, \gamma_1, \gamma_2)_N$). Values of $PoR < 1$ indicate that the new solution is more robust than the current schedule and $PoR > 1$ indicates that the new solution is less robust than the current one. Values of $PoR = 1$ indicates that the two solutions are identical in terms of robustness.

$$PoR = 100 \cdot \frac{F(\tau, \gamma_1, \gamma_2)_R}{F(\tau, \gamma_1, \gamma_2)_N} \qquad (2)$$

$$PoAD = 100 \cdot \frac{\sum_{m=1}^{M} \sum_{k=1}^{K} A_{m,k}}{K \cdot M} \qquad (3)$$

$$\begin{cases} A_{m,k} = 1 & if \ SD_{m,k} \geq \varepsilon \\ A_{m,k} = 0 & if \ SD_{m,k} < \varepsilon \end{cases} \qquad (4)$$

$$\begin{cases} A_{m,k} = 1 & if \ |SD_{m,k}| \leq \varepsilon \end{cases} \qquad (5)$$

The $PoAD$ indicator, shown in (3), returns the percentage of delays that a given schedule is able to absorb. According to (Naumann et al. 2011), the robustness of a bus schedule is related to its tolerance to delays, and thus this measure is considered utterly relevant in this context. In our empirical assessment of bus schedules, a delay is considered "absorbed" when its corresponding $SD_{m,k}$ is higher than a pre-defined threshold $\varepsilon > 0$, as shown in expression (4). Other measures of schedule performance could also be considered by changing the parameterization of the values of $A_{m,k}$. For example, the percentage of deviations could be measured using expression (5).

3 A Real World Case Study

Route 206 operates in Porto and has 6 TCPs. The total distance between TCP 1 and TCP 6 is 11.1 km, including 37 bus stops. This bus route is organized into five bus shifts on working days. The dataset of observations were gathered by the company "OPT - Optimização e Planeamento de Transportes SA", with resource to GPS devices available within all buses operating in Porto. The data covers a period of 3 weeks (13 working days). Figure 1 shows the daily observations for the first shift (gray lines) alongside with the current published schedule (black dashed line).

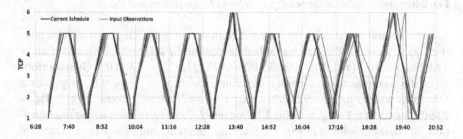

Fig. 1 Current schedule and the set of real observations of shift 1 of route 206.

Figure 1 evidences that shift 1 typically complies with schedule during the first half of the day, but tends to be delayed from 13h40 onwards. One of the observations incurred in a severe disruption. Buses on this shift tend to incur in anticipations from schedule only for the last run of the day.

4 Results and Discussion

The model was implemented in the IBM ILOG CPLEX Optimization Studio, version 12.6. The set of real data includes $K = 13$ daily observations. The daily trip covers a total of $M = 99$ TCPs. The total positions under assessment are 1287 (i.e. $K \cdot M = 13 * 99 = 1287$). All model runs were conducted keeping the current total trip time (Ψ) equal to 13 hours and 51 minutes.

A sensitivity analysis was conducted to study the influence of weights γ_1 and γ_2 in the robustness of the schedules returned by the model. All combinations of these parameters were tested considering that each of them could vary between 0 and 1 in intervals of 0.1. The schedules returned by the model were assessed using the measures PoR and $PoAD$, considering ε equal to 1 minute. The results are shown in Figure 2. Note that the current schedule has a PoR equal to 100%, and solutions are better than this reference when their values decrease.

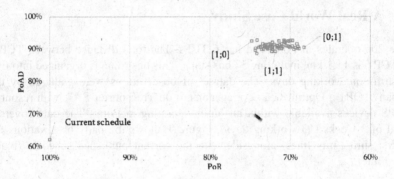

Fig. 2 Results obtained for the sensitivity analysis of weights γ_1 and γ_2.

For a neutral perspective on the relative importance of delays and anticipations (i.e., $\gamma_1 = \gamma_2 = 1$), the optimal schedule returned $PoAD =89.28\%$ and $PoR = 71.87\%$, where $F(\tau, 1, 1)_N = 3855.80$ and $F(\tau, 1, 1)_R = 2770.05$. Giving priority to the minimization of earlier departures, with $\gamma_1 = 1$; $\gamma_2 = 0$, originates a schedule with $PoAD = 92.38\%$ and $PoR = 74.78\%$, where $F(\tau, 1, 0)_N = 2150.65$ and $F(\tau, 1, 0)_R = 1608.32$. Conversely, giving priority to the minimization of delays, with $\gamma_1 = 0$; $\gamma_2 = 1$, the optimal solution has a value of $PoAD=90.44\%$ and $PoR=68.30\%$, where $F(\tau, 0, 1)_N = 1705.15$ and $F(\tau, 0, 1)_R = 1164.55$.

Figure 2 shows that all solutions obtained with the model are similarly robust and significantly more robust than the current schedule, considering their PoR and $PoAD$ values. The solutions returned by the model preferentially absorb delays and are not very sensitive to changes in γ_1 and γ_2, which is mainly related to the large amount of delays contained in observations. Figure 3 shows the dispersion of deviations from schedule in ranges of 5 minutes for the scenarios previously described.

Figure 3 shows the distribution of deviations for the schedules under analysis, considering all positions under assessment (i.e., K·M=1287). The distribution of deviations is similar for the three schedules obtained with the model. This was already expected from the previous analysis using robustness indicators. These schedules were able to significantly decrease delays, which implies an increase of positions where the bus is able to arrive earlier. The current schedule includes a similar number of delays and anticipations. Figure 4 provides the comparison between the current schedule (black dashed line) and the schedule obtained considering $\gamma_1=1$ and $\gamma_2=1$ (blue line).

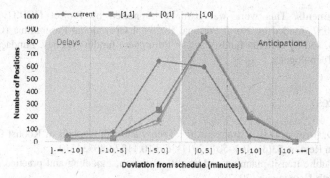

Fig. 3 Distribution of schedule deviations.

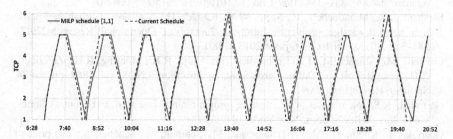

Fig. 4 Visualization of current schedule and new schedule.

Whilst in current schedule the slacks are incorporated exclusively in path inversion TCPs (i.e., TCP 1, 5 or 6), the schedules returned by the model provide the optimal re-distribution of slack times considering all TCP over the daily path.

5   Conclusions

This study applied an optimization model to improve robustness in bus schedules. The model returns the optimal allocation of slack time to incorporate in each position of the schedule considering a set of real observations. The model was applied to a real world case study. The robustness of the solutions obtained was assessed using the measures PoR and $PoAD$. The solutions obtained were also compared with the current schedule. A sensitivity analysis showed that the optimal solution preferentially absorb delays and is not very sensitive to the variation of weights γ_1 and γ_2, which is mainly associated with the large amount of delays comprised in the set of observations for the case study under analysis.

The implementation of schedules obtained using the techniques addressed in this study would endow buses with an adequate distribution of slack time, considering the range in which uncertainty fluctuates in observations. Therefore, buses would be expected to perform daily trips with a lower rate of delays. The implementation of new schedules should consider the results obtained from optimization techniques alongside with the opinion of managers of transportation services.

Acknowledgments. This work was partially supported by Project NORTE-07-0124-FEDER-000057, funded by North Portugal Regional Operational Programme (ON.2 - O Novo Norte), and by national funds, through Portuguese funding agency, Fundação para a Ciência e a Tecnologia.

References

Bertsimas, D., Brown, D.B., Caramanis, C.: Theory and Applications of Robust Optimization. Siam Review 53(3), 464–501 (2011). doi:10.1137/080734510

Ceder, A.: Public transit planning and operation: theory, modeling and practice. Elsevier, Butterworth-Heinemann (2007)

Ceder, A., Wilson, N.H.M.: Bus Network Design. Transportation Research Part B: Methodological 20(4), 331–344 (1986). doi:10.1016/0191-2615(86)90047-0

Farahani, R.Z., Miandoabchi, E., Szeto, W.Y., Rashidi, H.: A review of urban transportation network design problems. European Journal of Operational Research 229(2), 281–302 (2013). doi:10.1016/j.ejor.2013.01.001

Fischetti, M., Monaci, M.: Light robustness. In: Ahuja, R.K., Möhring, R.H., Zaroliagis, C.D. (eds.) Robust and Online Large-Scale Optimization. LNCS, vol. 5868, pp. 61–84. Springer, Heidelberg (2009)

Garfinkel, R.S., Nemhauser, G.L.: Integer programming. Decision and Control Series, vol. 4. John Wiley & Sons, New York (1972)

Goerigk, M., Knoth, M., Müller-Hannemann, M., Schmidt, M., Schöbel, A.: The price of robustness in timetable information. Paper presented at the ATMOS, – 11th Workshop on Algorithmic Approaches for Transportation Modelling, Optimization, and Systems, Saarbrücken, Germany, September 2011

Hill, S.A.: Numerical analysis of a time-headway bus route model. Physica A: Statistical Mechanics and Its Applications 328(1-2), 261–273 (2003). doi:10.1016/S0378-4371(03)00517-X

Ibarra-Rojas, O.J., Delgado, F., Giesen, R., Muñoz, J.C.: Planning, operation, and control of bus transport systems: A literature review. Transportation Research Part B: Methodological 77, 38–75 (2015). doi:10.1016/j.trb.2015.03.002

Kouvelis, P., Yu, G.: Robust Discrete Optimization and Its Applications. Nonconvex Optimization and Its Applications, vol. 14. Springer Science & Business Media, Netherlands (1997)

Liebchen, C., Lübbecke, M., Möhring, R., Stiller, S.: The concept of recoverable robustness, linear programming recovery, and railway applications. In: Ahuja, R.K., Möhring, R.H., Zaroliagis, C.D. (eds.) Robust and Online Large-Scale Optimization, pp. 1–27. Springer, Heidelberg (2009). doi:10.1007/978-3-642-05465-5_1

Liu, G., Wirasinghe, S.C.: A simulation model of reliable schedule design for a fixed transit route. Journal of Advanced Transportation 35(2), 145–174 (2001). doi:10.1002/atr.5670350206

Mulley, C., Ho, C.: Evaluating the impact of bus network planning changes in Sydney, Australia. Transport Policy 30, 13–25 (2013). doi:10.1016/j.tranpol.2013.07.003

Naumann, M., Suhl, L., Kramkowski, S.: A stochastic programming approach for robust vehicle scheduling in public bus transport. Procedia-Social and Behavioral Sciences 20, 826–835 (2011). doi:10.1016/j.sbspro.2011.08.091

Schmid, V.: Hybrid large neighborhood search for the bus rapid transit route design problem. European Journal of Operational Research 238(2), 427–437 (2014). doi:10.1016/j.ejor.2014.04.005

Szeto, W.Y., Wu, Y.Z.: A simultaneous bus route design and frequency setting problem for Tin Shui Wai, Hong Kong. European Journal of Operational Research 209(2), 141–155 (2011). doi:10.1016/j.ejor.2010.08.020

van Oudheusden, D.L., Zhu, W.: Trip frequency scheduling for bus route management in Bangkok. European Journal of Operational Research 83(3), 439–451 (1995). doi:10.1016/0377-2217(94)00362-G

Yan, Y.D., Meng, Q., Wang, S.A., Guo, X.C.: Robust optimization model of schedule design for a fixed bus route. Transportation Research Part C: Emerging Technologies 25, 113–121 (2012). doi:10.1016/j.trc.2012.05.006

Zhao, F., Zeng, X.G.: Optimization of transit route network, vehicle headways and timetables for large-scale transit networks. European Journal of Operational Research 186(2), 841–855 (2008). doi:10.1016/j.ejor.2007.02.005

Airport Ground Movement

Miriam Lobato[1], Filipe Carvalho[1], Ana Sofia Pereira[1], and Agostinho Agra[2]

[1] Wide Scope – Optimization Solutions, Lisboa, Portugal
[2] CIDMA and Departamento de Matemática, Universidade de Aveiro, Portugal
{miriam.lobato,filipe.carvalho,ana.pereira}@widescope.pt,
aagra@ua.pt

Abstract. Worldwide air traffic tends to increase and for many airports it is no longer an option to expand terminals and runways, so airports are trying to maximize their operational efficiency. Many airports already operate near their maximal capacity. Peak hours imply operational bottlenecks and cause chained delays across flights impacting passengers, airlines and airports. Therefore there is a need for the optimization of the ground movements at the airports. The ground movement problem consists of routing the departing planes from the gate to the runway for takeoff, and the arriving planes from the runway to the gate, and to schedule their movements. The main goal is to minimize the time spent by the planes during their ground movements while respecting all the rules established by the Advanced Surface Movement, Guidance and Control Systems of the International Civil Aviation. Each aircraft event (arrival or departing authorization) generates a new environment and therefore a new instance of the Ground Movement Problem. The optimization approach proposed is based on an Iterated Local Search and provides a fast heuristic solution for each real-time event generated instance granting all safety regulations. Preliminary computational results are reported for real data comparing the heuristic solutions with the solutions obtained using a mixed-integer programming approach.

1   Introduction

According to EUROCONTROL [5] air traffic is going to have a significant increase and due to this fast growth, airports have to maximize the use of their capacities. For many of them, it is no longer an option to expand terminals and runways. In order to satisfy the demand, airports must optimize their operations maintaining high security levels. Facing this new reality the International Civil Aviation Organization (ICAO) proposed an Advanced Surface Movement, Guidance and Control Systems (A-SMGCS).

Currently the Surface Movement, Guidance and Control Systems (SMGCS) procedures are based on the principle of "see and be seen". According to the manual the number of accidents during surface movements is increasing, because of air traffic growth; the increasing number of operations that take place in low

© Springer International Publishing Switzerland 2015
A.P.F.D. Barbosa Póvoa and J.L. de Miranda (eds.), *Operations Research and Big Data,*
Studies in Big Data 15, DOI: 10.1007/978-3-319-24154-8_11

visibility conditions; and the complexity of airports layout. Facing these problems ICAO proposed an upgrade to the system. The new manual specifies some system objectives and functions. An A-SMGCS should support some primary functions, for example, surveillance, routing, guidance and control. The same A-SMGCS should be capable of assisting authorized aircraft and vehicles to maneuver safely and efficiently on the movement area [8].

Given a set of departing aircrafts and a set of arriving aircrafts, the Ground Movement Problem (GMP) is to route and schedule all those aircrafts through the taxiways to their destination, in order to minimize the overall taxiing time, while satisfying a set of safety constraints and operational rules.

From the departing aircraft's point of view, the goal is to reach the runway as soon as possible. For the arriving aircrafts the problem only begins when they reach the holding point that gives access to the taxiway. The goal is to reach the gate as soon as possible. These goals also contribute to the minimization of the fuel consumption and minimization of the time spent on the time window slot for occupying the airport ground. Also passengers and the environment benefit from an optimized ground movement.

Safety regulations impose constraints such as time separation between aircrafts due to jet blast, and aircraft movement speed limits. The sequence of airplanes to use the runways and the corresponding time schedule is pre-defined.

The problem starts either when an aircraft is authorized to move from its parking position to a position where it can start taxiing, or when it lands at the runway. Therefore the problem will be solved several times within an hour and should be solved very quickly (in few seconds). In each problem the starting position and the destination of each airplane is known.

The GMP can be easily solved for low-activity airports or for low-activity periods. But that is not what happens for most international airports where during peak periods there are many aircrafts moving from and to the runways. Typically the current practical approach is to put the planes in queues in order to minimize the number of conflicts between aircrafts. This has the disadvantage of increasing the taxi time unnecessarily.

Variants of the GMP have been studied before. In 2001 Pesic *et al.* presented a Genetic Algorithm and report computational tests on the Charles De Gaulle Airport. Later, Smelting *et al.* (2004) introduce a mixed-integer programming (MIP) formulation. They assume the routes are given and an arrival or departure time for each aircraft is given as well. The decision relays on the time that each airplane is going to leave a particular point at the airport such that no conflicts occur, and all airplanes meet the time requirements. Three different variants of rolling horizon algorithms were implemented. Real data from Amsterdam Airport Schipol was used to demonstrate that the algorithms lead to significant improvements of efficiency, with reasonable computational effort. A new and more complex MIP model was introduced by Keith and Richards (2008) to optimize runway and taxiway operations together. Their approach was not applied to real data. They only replicate London Heathrow Airport's north runway east holding point structure. Roling and Visser (2008) revisited the work developed by Smelting et al. [13] and presented a new MIP formulation based on a discretized time horizon. The major difference between these two works is that in [13] range speeds are allowed but holding or rerouting are not permitted.

Atkin *et al.* (2010) presented a survey on the GMP. There MIP models and heuristics are reviewed. The authors claim that MIP approaches are not capable of finding solutions in a reasonable time, so heuristic methods should be applied. For heuristics only Genetic Algorithms are mentioned.

2   Mathematical Formulation

The ground movements of planes at airports can be modeled as flows on a directed graph, $G=(V,A)$, defined by the airport ground layout, where V is the set of nodes, which represent the initial aircraft positions; holding points at intersections; holding positions of taxiways; intersection of taxiways; and the runway. Set A is the set of arcs, representing the existing connections between pairs of nodes. Similarly to the formulation given in [12], the model presented here considers a discrete time representation of the planning horizon. The model considers an expanded network where a copy of each node in V is created for each time period.

The set of time periods is represented by $T = \{1, ..., |T|\}$. In each time period an aircraft can be taxiing or waiting. An aircraft can wait at the origin, or at each node, preventing conflicts from happening (for example having two aircrafts reaching the same node or having two aircrafts disrespecting the safety margin).

In order to guarantee feasibility, delays for takeoffs are allowed and a penalty β is associated to those delays.

A set $P = \{1, ..., |P|\}$ of airplanes in the system is considered. The set of planes is composed by two subsets, P^D representing departing aircrafts and P^A representing arriving airplanes. $P^D \cup P^A = P$. The set of different categories of airplanes is denoted by $C = \{1, ..., |C|\}$. For each plane the following parameters are considered: o_p is the initial position (origin) of aircraft p; d_p is the destination of aircraft p; a_p is the predecessor of plane p in the runway ($a_p = 0$ if there is no antecessor); PT_p is the scheduled time for airplane $p \in P$ to use the runway (it also defines the runway sequence implicitly); ST_p is the time at which airplane $p \in P^D$ is available at origin; PC_p represents the airplane type/category and sp_p is the separation time between two consecutive planes when plane p is ahead (depends only on the category of airplane p). Additionally k_{ijp} is the taxi time for aircraft p to go from i to j, $(i, j) \in A$;

Furthermore, we define the following binary variables: x_{ijpt} $p \in P$, $(i,j) \in A$, $t \in T$ is 1 if aircraft p passes through arc $(i,j) \in A$ at period t and zero otherwise; y_{jpt} $j \in V$, $p \in P$, $t \in T$ is 1 if aircraft p holds in position $j \in V$ at period t and zero otherwise, and z_{itc}, $i \in V$, $t \in T$, $c \in C$ is 1 if an aircraft of type c is in position i at period t and zero otherwise. All the variables that do not exit, such as x_{ijpt} and y_{jpt} with $t < ST_p$ are assumed to be zero in the model.

The mathematical formulation is as it follows:

$$Min \quad \sum_{p \in P^D} \beta_p \left(\left(\sum_{i \in V} \sum_{t \in T | t > k_{i,d_p,p}} tx_{i,d_p,pt-k_{d_p p}} \right) - PT_p \right)$$

$$+ \sum_{i \in V} \sum_{j \in V} \sum_{p \in P} \sum_{t \in T} x_{ijpt} + \sum_{j \in V} \sum_{p \in P} \sum_{t \in T} y_{jpt} \tag{1}$$

$$\sum_{i \in V} x_{o_p,ip,ST_p} + y_{o_p,ip,ST_p} = 1 \qquad\qquad p \in P \tag{2}$$

$$\sum_{i \in V} \sum_{t \in T} x_{i,d_p,pt} = 1 \qquad\qquad p \in P \tag{3}$$

$$\sum_{i \in V} x_{ijp,t-k_{ijp}} + y_{jp,t-1} = \sum_{i \in V} x_{jipt} + y_{jpt} \qquad \begin{array}{l} j \in V, p \in P, t \in T, \\ t > ST_p, j \neq d_p \end{array} \tag{4}$$

$$\sum_{i \in V} \sum_{t \in T} tx_{i,d_p,p,t-k_{id_p p}} \geq \sum_{i \in V} \sum_{t \in T} tx_{i,d_{a_p},a_p,k_{i,d_{a_p}a_p}} \qquad \begin{array}{l} p \in P^D, a_p \in P^D, \\ a_p > 0, \end{array} \tag{5}$$

$$\sum_{i \in V} \sum_{t \in T} tx_{i,d_p,p,t-k_{i,d_p,p}} \geq \sum_{i \in V} \sum_{t \in T} tx_{o_{a_p},i,a_p,t} \qquad \begin{array}{l} p \in P^D, a_p \in P^A, \\ a_p > 0 \end{array} \tag{6}$$

$$\sum_{p \in P} \sum_{i \in V} \sum_{l \in \{t,...,t+sp_c-2\}} x_{ijp,l-k_{ijp}} \leq z_{jtc} + sp_c \left(1 - z_{jtc}\right) \qquad c \in C, t \in T, j \in V \tag{7}$$

$$\sum_{c \in C} z_{itc} \leq 1 \qquad\qquad i \in V, t \in T \tag{8}$$

$$\sum_{p \in PC_c} \sum_{i \in V} x_{ijp,t-k_{ijp}} + \sum_{p \in PC_c} y_{jp,t-1} = z_{jtc} \qquad j \in V, t \in T, c \in C \tag{9}$$

$$x_{jipt}, y_{jpt}, z_{itc} \qquad \text{are binary} \tag{10}$$

The objective function (1) is to minimize the penalty for delays for takeoffs (first term) plus the taxiing time (second and third terms). This objective function is a utility function that aggregates two different types of objectives: minimize the delays and minimize the taxiing time. This practical approach is motivated by the need to follow a fast solution procedure that forces the decision maker to take his/her decisions *a priori* (in this case that can be accomplished by adjusting the penalty parameter). The components are weighed by parameter β, the penalty for takeoff delays. Constraints (2), (3) and (4) model a flow for each plane. Constraints (2) determine that each airplane either waits at its origin or starts taxiing. Constraints (3) assure that each airplane always reaches its destination. Constraints (4) guarantee flow conservation at each expanded node. Constraints (5) and (6) ensure that a plane can only use the runway after its antecessor. Constraints (7) guarantee that two aircraft maintain a safety margin depending on the type of airplane that goes ahead. Constraints (8) ensure that at most one airplane is assigned to an expanded node while Constraints (9) state that an expanded node (j,t) is assigned to a plane type c if one of plane of that class arrives at node j in period t or it has been waiting from the previous period. Constraints (10) are the sign constraints.

Although the model uses a discrete time as in [12] the two formulations are quite different. For instance, in [12] a path-type formulation is considered, that is, the variables are associated to routes and not to arcs as in here. Also the type of planes is ignored in [12].

3   Heuristics

The airport ground movement problem as well as the discussed variants is a NP-Hard problem [7]. The size of the problem instances makes it very difficult to find an optimal solution through exact methods within an acceptable runtime for airports peak hours. Thus heuristic approaches were developed and the quality of the heuristic solutions was assessed by comparing it against the optimal solutions obtained by the exact approach (Branch and Cut).

The heuristic schemes include three steps: (i) preprocessing; (ii) construction of a feasible solution; (iii) improving the solution through a local search.

Preprocessing
k-shortest paths between every two nodes are computed and stored. This computation was obtained through the implementation of a Depth-First Search Algorithm (DFS) [2]. Still in the preprocessing of data, the sequence of planes using the runway is determined taking into account the time each plane has to use it.

Constructive algorithms
The initial solution is obtained through a constructive algorithm. For this purpose, a greedy heuristic and a random heuristic were implemented. The greedy heuristic assigns to each plane the shortest path (the one with less taxi time) between an origin and a destination. The other constructive heuristic chooses the path random-

ly between the k-shortest paths. Then, in both heuristics, the planes are sorted accordingly to the time they use the runway(s). Following that order the schedule of moves of each plane is computed in order to minimize the arrival time to its destination. Notice that once a schedule is computed for a plane, that schedule is fixed during the constructive phase. This schedule may create conflicts with other planes that are to be scheduled. If there is a conflict at a given node, the last plane to be scheduled among those in conflict must wait. Notice that if there are no conflicts, as in the case of many low activity periods, the greedy heuristic leads to the optimal solution.

Improving heuristic

The Iterated Local Search (ILS) meta-heuristic was implemented in order to escape local optima, by diversifying and intensifying search during the available computation time. The available computation time is in fact a short time window (pointed by the operation to be desirably lower than 10 seconds)

Following the order of planes, for each plane, the procedure checks whether there is an improvement by replacing the current path by each other of the k-shortest paths. Two different processes of local search were implemented, First Ascent and Steepest Ascent. The First Ascent method searches for the first better neighbor. This search is made by changing the route, for one plane at a time. When a route that improves the current solution is found the search for that plane stops. And so on for the other planes. The Steepest Ascent method searches through every neighbor, following some predefined criteria, and chooses the best one. The Steepest Ascent searches, in plane one, all routes, one by one, choosing the route that improves solution. If there is no route that improves solution, the current solution is kept. And so on for the other planes. Thus while First Ascent stops in the first better neighbor, Steepest Ascent keeps searching despite having found a better solution already. Computational tests show that the First Ascent approach produces solutions with the same accuracy, in a faster way.

The Iterated Local Search method uses the constructive heuristic with random paths, and then searches the described neighborhood to find better solutions. In other words, the random search is performed and then a local search is made based on the solution found.

4   Computational Results

The computational tests were conducted with real data from an airport. The tests were run on a PC with an Intel(R) core (TM) 2 duo CPU T8300 @ 2.4 GHZ.

From the observation of results it was possible to conclude that using the shortest paths was always better than using randomly chosen paths, despite the difference is not that big.

Next we present the running times, in seconds, obtained when solving 11 instances with the ILS, assuming time horizons of 1 hour each.

Hour	8h	9h	10h	11h	12h	13h	15h	16h	17h	18h	19h
Time	1.2	1.06	1.02	1.03	1.02	1.03	1.04	1.09	1.05	1.05	1.03

The table shows that for very large time horizons of one hour ILS is very fast. Next ILS is compared with the Branch and Cut (BC), using Xpress-Optimizer. Given limitations of the exact approach, in what concerns runtime and memory, the problem was divided into time windows of 15 minutes. Only the peak hours where chosen. The peak hour in the morning is from 8h to 9h, and in the afternoon the peak hour is from 19h to 20h. In 8 given problem instances, ILS always found the same solution as BC, i.e., it could be observed that the paths followed by planes were the same. With such time windows the number of planes considered in each iteration is not high. Among the tested instances the largest number of airplanes considered was 11. Thus the number of conflicts obtained by applying the constructive heuristic is very small which justifies the very good results obtained by the ILS. The following table presents the running times of both algorithms.

Time Frames	Branch and Cut	Iterated Local Search
8h00 – 8h14min	481s	0.16s
8h15min – 8h29min	248s	0.11s
8h30min – 8h44min	770s	0.13s
8h45min – 8h59min	662s	0.11s
19h00 – 19h14min	445s	0.12s
19h15min – 19h29min	53s	0.07s
19h30min – 19h44min	169s	0.09s
19h45min – 19h59min	262s	0.11s

As depicted the same solutions are found for the most relevant peak time frames by both approaches. However, the ILS approach is capable of determining solutions much faster. Please keep in mind that a new solution is ought to be produced at every plane "touch-ground event" before it reaches the first stop bar right after the runway.

5   Conclusions and Future Research

An optimization procedure was proposed to solve the Ground Movement Problem. This scheme proved to be very efficient to solve real instances from the considered airport during peak hours. The heuristic scheme is very fast and obtained the optimal solution for all tested instances considering a time horizon of 15 minutes, which corresponds to the real size of the planning horizons.

Further work will include comparing the savings of using this optimization method with the real performance of human decisions being taken nowadays.

Also, further work shall be developed for scenarios with higher simultaneous traffic while trying to improve computation times even further.

Acknowledgements. The research of the fourth author was partially supported by Portuguese funds through the *Center for Research and Development in Mathematics and Applications* (CIDMA) and FCT, the Portuguese Foundation for Science and Technology, within project UID/MAT/04106/2015.

Wide Scope research was co-funded by the European Regional Development Fund (ERDF), via the National Strategic Reference Framework (NSRF) and contracted by Instituto de Apoio às Pequenas e Médias Empresas e à Inovação (IAPMEI), through the contract nr. 2013/30149.

References

[1] Atkin, J.A.D., Burke, E.K., Ravizza, S.: The airport ground movement problem: paste and current research and future directions. In: 4th International Conference on Research in Air Transportation (2010)

[2] Cardoso, D., Szymanski, J., Rostami, M.: Matematica Discreta. Escolar Editora (2009)

[3] Clare, G., Richards, A.: Receding Horizon Iterative Optimization of Taxiway Routing and Runway Scheduling. American Institute of Aeronautics and Astronautics (2009)

[4] Cambridge Professional English, Flightpath. Aviation English for Pilots and ATCOs - Glossary of Aviation Terms. Cambridge University Press (2011)

[5] European Organization for the Safety of Air Navigation (EUROCONTROL). EUROCONTROL Seven-Year Forecast (2014)

[6] Glover, F., Laguna, M.: Tabu Search. Kluwer Academic Publishers (1997)

[7] Godbole, P.J., Ranade, A.G., Pant, R.S.: Routing and Scheduling Algorithm for Aircraft Ground Movement Optimization. American Institute of Aeronautics and Astronautics

[8] International Civil Aviation Organization (ICAO). Advanced Surface Movement Guidance and Control Systems (A-SMGCS) Manual, 1st edn. (2004)

[9] Keith, G., Richards, A.: Optimization of taxi routing and runway scheduling. In: Proceeding of the AIAA Guidance, Navigation and Control Conference, Honolulu, USA (2008)

[10] Marin, A., Codina, E.: Network Design - Taxi Planning (2008)

[11] Pesic, B., Durand, N., Alliot, J.: GECCO 2001 Conference: Real-World Applications Aircraft Ground Traffic Optimization using a Genetic Algorithm (2001)

[12] Roling, P.C., Visser, H.G.: Optimal airport surface traffic planning using mixed-integer linear programming. International Journal of Aerospace Engineering 2008(1), 1–11 (2008)

[13] Smeltink, J.W., Soomer, M.J., Wall, P.R., de, M.R.D.: van der, An optimization model for airport taxi scheduling. In: Proceedings of the INFORMS Annual Meeting, Denver, USA (2004)

Optimization of a Recyclable Waste Collection System – The Valorsul Case Study

Diogo Lopes[1], Tânia Rodrigues Pereira Ramos[2], and Ana Paula Barbosa-Póvoa[1]

[1] CEG-IST, Instituto Superior Técnico, Universidade de Lisboa, Lisbon, Portugal
[2] Instituto Universitário de Lisboa (ISCTE-IUL),
  Business Research Unit (BRU), Lisbon, Portugal

Abstract. This paper studies alternative scenarios for a recyclable waste collection system in order to increase efficiency in their operations. Three alternative scenarios are proposed where two different locations for one or two additional depots are studied. The problem is considered as a multi-depot vehicle routing problem and a solution approach is developed. The three scenarios are compared with the current solution regarding distance travelled, working hours, amount of waste collected and vehicle usage. Significant gains are obtained when depots are added to the current logistics system.

1   Introduction

Nowadays, organizations are faced with an extremely competitive environment, which triggers off a search process for efficiency and effectiveness in its processes and operations. Academic research has a role to play in this context as it can be very helpful to reach this purpose allowing scientifically supported analysis of possible alternative scenarios to the current operation.

This paper aims to study the current operation of a Portuguese recyclable waste collection company – Valorsul, and proposes and analyses alternative scenarios to improve Valorsul's efficiency. The recent decrease in waste production, the increasing awareness with environmental issues and the need of compliance with the recycling targets imposed by the European Union had motived this kind of studies in companies responsible for the collection of recyclable waste. Such factors enhance the need of making thoughtful decisions that reduce costs and diminish the resources required for the operation. Given that the collection cost, specially the fuel cost, is of great importance in the cost structure of the company, any reduction in the distance travelled in the collection activity will have an immediate impact in the total cost and will improve the company's performance.

Valorsul is responsible for the selective collection in 14 municipalities located in Portugal West Region, covering about 3000 km^2. One depot and six transfer stations are owned by the company. All vehicles start and end the collection routes at the depot. Long routes are performed given its vast operating service area. There is then a potential space for optimization, where vehicle routes could be

© Springer International Publishing Switzerland 2015   97
A.P.F.D. Barbosa Póvoa and J.L. de Miranda (eds.), *Operations Research and Big Data*,
Studies in Big Data 15, DOI: 10.1007/978-3-319-24154-8_12

optimized through the reduction of the allocated service area. As suggested by the company, the existent transfer stations could turn easily into depots. For that, is only needed to split the actual fleet between the actual depot and the new ones. The number and which transfer stations would turn into depots are the questions that will be explored in this paper.

The paper is structured as follows: in section 2, the Multi-Depot Vehicle Routing Problem (MDVRP) and its application to waste collection systems are explored. In section 3, the main problem characteristics are presented. In section 4 the solution approach is characterised. In section 5 a set of alternative scenarios is solved and the results discussed. Lastly, in section 6 the conclusions are presented.

2   Literature Review

2.1   MDVRP

MDVRP is a generalization of the vehicle routing problem (VRP) in which, beyond the definition of vehicle routes, it is necessary to determine from which depot customers are to be visited. Over the years, several MDVRP models have been developed. These models allow finding exact and approximate solutions. However, since it is a NP-hard problem, the existing models in the literature are mostly heuristic-based. Tillman and Cain (1972) developed a heuristic based on the savings method, modifying the distance formula to enable the existence of multiple depots. Some years later, Golden et al. (1977) proposed two heuristic algorithms which allow solving larger problems. In the first algorithm customers are assigned to depots while the routes are defined. In the second algorithm, the customers are first assigned to depots and then, in a second phase, the routes are defined through a heuristic algorithm for the VRP. Other heuristics have been proposed to solve MDVRP, including those proposed by Renaud et al. (1996) and Crevier et al. (2007). On the exact models, Laporte et al. (1984) and Laporte et al. (1988) developed exact branch and bound algorithms to solve symmetric and asymmetric versions of the MDVRP, but which are only applicable to small instances. More recently, in 2009, Baldacci and Mingozzi (2009) developed an exact method, which is able to solve, among others, the MDVRP.

2.2   Applications to Waste Collection Systems

Regarding waste collection systems, some works have been published in the literature in the recent years (an extensive survey can be found in the work of Ghiani et al., 2014). Tung and Pinnoi (2000) present a heuristic for a problem of establishing routes and scheduling a fleet of vehicles with multiple time windows. Angelelli and Speranza (2002) developed a model for solving the periodic vehicle routing problem (PVRP), which is applicable to different waste collection systems. Using a tabu search algorithm, the model was applied to an undifferentiated waste collection system in Italy and to a system for collecting paper and organic waste in Belgium. Teixeira et al. (2004) present a heuristic approach divided in three phases to solve a PVRP. The three phases are solved through heuristics. This approach was applied to a real case in Portugal. Recently, Ramos et al. (2014) studied the problem of planning the collection of recyclable waste taking into account economic and environmental factors. In this work, a Multi-Product, Multi-Depot Vehicle Routing Problem (MP-MDVRP) is tackled, where the service areas and vehicle routes have to be defined in a logistics network with multiple depots and multiple products to be collected. A Mixed-Integer Linear Programming (MILP) model is proposed for the MP-MDVRP and a decomposition approach is developed and applied to a recyclable waste collection system operating in Portugal. The results were the reduction of the distance travelled and the reduction of CO_2 emissions, thus ensuring an improvement both economically and environmentally.

3   Case-Study

Valorsul is responsible for the collection of 7807 containers (2542 paper containers, 2378 plastic/metal containers and 2887 glass containers) located at 14 municipalities. Valorsul owns a homogenous vehicle fleet of 12 vehicles that is based only at one depot, located in *Centro de Tratamento de Resíduos do Oeste* (CTRO), in the municipality of Cadaval. Besides this facility, Valorsul owns six transfer stations located at Nazaré, Óbidos, Peniche, Rio Maior, Alenquer and Sobral de Monte Agraço. Currently, there are 82 routes established, 26 for paper collection, 26 for plastic/metal and 30 for glass. Considering the routes performed between January and September 2013, Table 1 shows the average collection frequency, number of containers collected by route, distance travelled per route and amount of material collected per route for each type of recyclable material. Paper and plastic/metal have similar indicators. The average distance travelled per route is about 136 km and each container is collected, on average, every 9 or 8 days. Glass has the highest time interval between consecutive collections (20,5 days) and longer routes (151 km), with a similar number of containers collected.

Table 1 Current indicators for Valorsul routes performed between January and September 2013 (average values)

Recyclable material	No. routes performed	Collection frequency (days)	No. containers collected per route	Distance travelled per route (km)	Amount collected per route (ton)
Paper	816	9,3	82	135,3	2.8
Plastic/Metal	740	8,3	81	137,9	2.1
Glass	342	20,5	83	151,2	9.8

After the analysis of the operation, several constraints that hinder the operation were identified. The size of the service area and also its asymmetry to CTRO location appear as the most important. The time limit of a shift, the location of containers in urban areas and its difficult access, and uncertainty about the filling levels were also identified as factors that could be further studied to improve the operation. However, this work will focus on the problems regarding the size and assymetry of the actual service area. The alternatives identified as the most suitable to mitigate those problems were to add more depots to the system, using the existent transfer stations as depots, i.e., where vehicles would be based to start and end the collection routes. Given the actual asymmetry of the CTRO location, the new potential depots should be located at the north of the actual service area. Therefore, three alternative scenarios were defined considering the existent facilities and the company's established options. Scenarios I and II study a solution with two depots: CTRO and Nazaré and CTRO and Óbidos. Scenario III considers a solution with three depots, located at CTRO, Nazaré and Óbidos (see Figure 1). It is expected that these scenarios will reduce the route duration and distance travelled. Each scenario will be evaluated and compared with the current solution in order to assess the best alternative to be implemented.

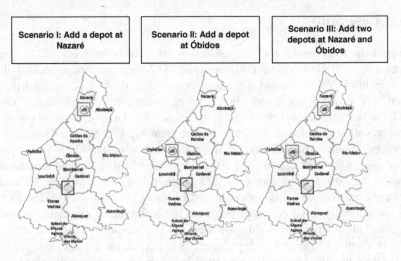

Fig. 1 Proposed scenarios

4   Solution Approach

Given the size of the problem (7807 containers grouped into 140 clusters), it is intractable to solve it with an exact algorithm (Ramos et al. 2014). Therefore, a decomposition approach has been developed based on the work of Ramos et al. (2014), where some constraints of the original problem are firstly removed and the relaxed problem is solved. Moreover, each material is solved separately as different service areas for each material are allowed by the company. The solution approach is illustrated at Figure 2. Firstly, the route duration constraints and the definition of only closed routes are removed. Then, the solution obtained is analysed and the feasible routes are separated from the unfeasible ones. A feasible route is a closed route with duration less or equal to the maximum time allowed for a route (450 minutes). The duration of a route includes the time to collect each container and the travel time. An unfeasible route is an open route or a closed route with duration superior to the time limit. For the collection sites that belong to the unfeasible routes, a second step is performed where a problem with duration constraints is solved. Again, the solution is analysed and the feasible routes are separated from the unfeasible ones. For the latter, a step 3 is performed where the original problem is solved, that is, a MDVRP is solved, but only for the sites that belong to the unfeasible routes. In step 1, a MDVRP with Mixed Closed and Open Inter-Depot Routes (MDVRP-MCO) is solved through the model proposed by Ramos et al. (2013), where constraints regarding route duration were removed. In step 2, the same model is used, but with duration constraints. In step 3, a MDVRP is solved through the model proposed by Ramos et al. (2014). The mathematical formulations used in all problems were based on the Two Commodity Flow Formulation proposed by Baldacci et al. (2004). This solution approach is applied to each recyclable material separately (paper, plastic/metal and glass).

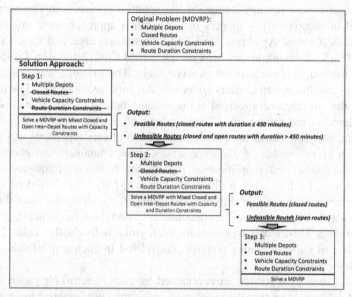

Fig. 2 Solution approach developed

The solution approach goal is to define vehicle routes that minimize the monthly distance travelled in the collection activity, for each recyclable material. The distance travelled also includes the waste transportation from the depots to the sorting station. The problem can be summarized as:

Given:

- Location of depots, sorting stations and collection sites;
- Distance between the different pairs of entities;
- Amount of material to collect at each collection site;
- Collection frequency for each collection site;
- Vehicle capacity to collect each material and vehicle speed;
- Time required to collect a container and to unload a vehicle;

Determine:

- The collection routes and its duration;
- The number of collection sites covered by each route;
- The amount of material collected on each route;

So as to minimize the total monthly distance and ensure total waste collection.

5   Results

This section covers the application of the solution approach to the case study described in section 3. The models in the solution approach were implemented with GAMS (General Algebraic Modeling System) language, and solved through CPLEX (23.5.1 version), in an Intel Core i3-2310M CPU, 2.10 GHz. To apply the solution method, two simplifications were made. The first one was to group the individual containers into clusters so as to reduce the problem size. For that, we consider that all containers located at one locality belong to the same cluster and are collected at the same route. Thus, instead of dealing with 7807 individual containers, we group them into 140 clusters. Each cluster represents a collection site with a given number of glass containers, paper containers and plastic/metal containers. The second simplification regards the collection frequency of each collection site. Each site has a different collection frequency. Based on the real operation we considered four different collection frequencies (one, two, three and four times a month) and group the collection sites into these frequencies, given the historical data. The models were run for each group individually. Table 2 shows the number of collection sites that was considered in each step of the solution approach for Scenario I.

Figure 3 shows the service areas obtained for each scenario for paper. A more balanced solution regarding the number of collection sites assigned to each depot is accomplished with scenario II. In Scenario III, depot Nazaré has few collection sites assigned to it.

In order to compare the scenarios three key performance indicators (KPI) were considered: vehicle usage rate, amount of material collected per kilometer and amount of material collected per minute. Figure 4 shows the comparison between the three scenarios and the current solution for the amount of material collected per kilometer and per minute indicators.

Table 2 Number of collection sites considered along the solution approach for Scenario I

Recyclable Mat.	Paper		Plastic/Metal		Glass	
Collection Freq.	4x/month	2x/month	4x/month	2x/month	2x/month	1x/month
Step 1: Input	83	56	112	26	5	135
Step 1: Output						
- No. collection sites belong to feasible routes	14	0	14	0	5	59
- No. collection sites belong to unfeasible routes	69	56	98	26	0	76
Step 2: Input	69	56	98	26	-	76
Step 2: Output						
- No. collection sites belong to feasible routes	60	56	79	26	0	76
- No. collection sites belong to unfeasible routes	9	0	19	0	0	0
Step 3: Input	9	-	19	-	-	-

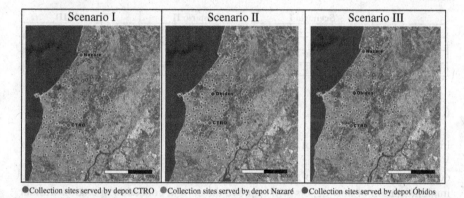

| Scenario I | Scenario II | Scenario III |

●Collection sites served by depot CTRO   ●Collection sites served by depot Nazaré   ●Collection sites served by depot Óbidos

Fig. 3 Service areas for paper in each scenario

Regarding the first indicator, scenario I reveals an improvement only for paper. On the other hand, scenarios II and III lead to improvements for paper and glass and maintains the values for plastic/metal when compared to the current scenario. For the second indicator, it suffers an improvement in all alternative scenarios and for all materials.

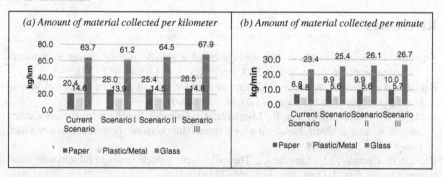

Fig. 4 Amount of material collected per kilometer and per minute for each scenario

In order to assess which is the best alternative scenario, the percentage between each scenario and the current solution for the KPIs was computed (see Table 3). Paper suffers the greatest improvement in all KPIs for all scenarios. Scenario I leads to a total improvement of 2.4% regarding kg/km and 22% regarding kg/min. Scenario III increases those results to 9.3% and 24%, respectively. Therefore, Scenario III, where two more depots are added to the current network, is the best scenario amongst the three. However, adding two more depots could be more difficult to implement than adding just one depot. If Valorsul concludes that the complexity inherent to the implementation of two more depots in relation to just one depot is not worth the gains, then Scenario II should be implemented (adding a depot at Óbidos).

Table 3 Comparison of the improvement % between alternative scenarios and current scenario

	Scenario I			Scenario II			Scenario III		
KPIs	Paper	Plastic	Glass	Paper	Plastic	Glass	Paper	Plastic	Glass
Vehicle usage	29%	3%	-1%	21%	3%	8%	21%	0%	1%
kg/km	23%	-5%	-4%	25%	-1%	1%	30%	1%	7%
kg/min	46%	17%	9%	46%	17%	12%	47%	19%	14%

6   Conclusions

In this paper a new way of managing the recyclable waste collection system of Valorsul is analysed. The problem was considered as a MDVRP and a decomposition technique to solve it was developed. The results indicate that the implementation of new depots will improve the current collection operation without the need of investment, given that existent facilities can be used as depots.

As future work the studied problem should be analysed by extending the used model to a location-routing problem where aspects such as: lengthening the shift duration, using open inter-depot routes in the operation, using a more accurate method to estimate the containers filling level and allowing different collection frequencies should be studied.

References

Angelelli, E., Speranza, M.G.: The application of a vehicle routing model to a waste-collection problem: two case studies. J. Oper. Res. Soc. 53, 944–952 (2002)

Baldacci, R., Mingozzi, A.: A unified exact method for solving different classes of vehicle routing problems. Math. Program. 120, 347–380 (2009)

Baldacci, R., Hadjiconstantinou, E., Mingozzi, A.: An exact algorithm for the capacitated vehicle routing problem based on a two-commodity network flow formulation. Oper. Res. 52, 723–738 (2004)

Crevier, B., Cordeau, J.F., Laporte, G.: The multi-depot vehicle routing problem with inter-depot routes. Eur. J. Oper. Res. 176, 756–773 (2007)

Ghiani, G., Laganà, D., Manni, E., Musmanno, R., Vigo, D.: Operations research in solid waste management: a survey of strategic and tactical issues. Comput. Oper. Res. 44, 22–32 (2014)

Golden, B.L., Magnanti, T.L., Nguyen, H.Q.: Implementing vehicle routing algorithms. Networks 7(2), 113–148 (1977)

Laporte, G., Nobert, Y., Arpin, D.: Capacitated multi-depot vehicle routing problems. Congr. Numer. 44, 283–292 (1984)

Laporte, G., Mercure, H., Nobert, Y.: An exact algorithm for the asymmetrical capacitated vehicle routing problem. Networks 16(1), 33–46 (1986)

Renaud, J., Laporte, G., Boctor, F.F.: A tabu search heuristic for the multi-depot vehicle routing problem. Comput. Oper. Res. 23, 229–235 (1996)

Ramos, T.R.P., Gomes, M.I., Barbosa-Póvoa, A.P.: Planning waste cooking oil collection systems. Waste Manage. 33, 1691–1703 (2013)

Ramos, T.R.P., Gomes, M.I., Barbosa-Póvoa, A.P.: Economic and environmental concerns in planning recyclable waste collection systems. Transport. Res. E-Log. 62, 34–54 (2014)

Teixeira, J., Antunes, A.P., de Sousa, J.P.: Recyclable waste collection planning – a case study. Eur. J. Oper. Res. 158, 543–554 (2004)

Tillman, F.A., Cain, T.M.: An upperbound algorithm for the single and multiple terminal delivery problem. Manage. Science 18(11), 664–682 (1972)

Tung, D.V., Pinnoi, A.: Vehicle routing-scheduling for waste collection in Hanoi. Eur. J. Oper. Res. 125, 449–468 (2000)

Simulated Annealing for Production Scheduling: A Case Study

António Santos Marques[1], Nelson Chibeles-Martins[2], and Tânia Pinto-Varela[1]

[1] CEG-IST, Instituto Superior Técnico, Universidade de Lisboa, A. Rovisco Pais, 1049-001 Lisboa, Portugal
[2] Centro de Matemática e Aplicações, Faculdade de Ciências e Tecnologia, Universidade Nova de Lisboa, Qta da Torre, 2829-516 Caparica, Portugal

Abstract. In this work, a production scheduling problem is developed through a Simulated Annealing approach, with the Tardiness minimization as objective function. The algorithm is applied to a plastic container facility, which operates an injection molding process with several parallel machines. The algorithm applicability and its performance is illustrated through two production strategies, the make-to-order (MTO) and make-to-stock (MTS). In the latter inventory management is taken into account.

1 Introduction

Nowadays the uncertainty, competitiveness and globalization has huge impact on all manufactures facilities and their markets. The demand uncertainty triggers high prices variability not only in the final product but also in the raw materials. This environment prompts the performance improvement of all companies' process, in order to reduce their operating costs.

On the other hand, the new technologies and software developments allow sophisticated operations' management enhancing production capabilities and efficiency. Therefore companies are investing time and money on the development of customized information systems not only to manage, plan and schedule its manufacturing process, but also control de inventory and its distribution.

When the process efficiency and productivity improvement is the goal, a support decision system to characterize an efficient scheduling in a reasonable time is very important.

Despite some research has already been made in this area using exact approaches, recently development have been also focused on meta-heuristics.

Chibeles-Martins et al. [1] developed Simulated Annealing methodology for the optimal design and scheduling of multipurpose batch plants and its comparison was made with the exact approach. The same authors [2], in the following year extended the previous work through the application of a multi-objective approach, and an efficient frontier characterization as a decision support tool was developed.

A.P.F.D. Barbosa Póvoa and J.L. de Miranda (eds.), *Operations Research and Big Data,*
Studies in Big Data 15, DOI: 10.1007/978-3-319-24154-8_13

Roshanaei et al. [3] developed a methodology for the flexible job shop scheduling problem through a SA algorithm, for large instances. Jia et al. [4] considered a more generic scheduling algorithm, for n jobs, with variable lots dimensions, using m identical and parallel batch machines, in order to minimize the makespan. Shiva-sankaran et al. [5] developed a hybrid sorting immune simulated annealing technique for solving a multi-objective flexible job-shop scheduling problem.

In this paper a Simulated Annealing algorithm is developed for an injection molding process schedule. The algorithm performance is illustrated in a real case study through the characterization of two process strategies and its statistical results analysed.

2   Modelling Framework

The meta-heuristic approach developed in this work in based in the Simulated Annealing algorithm proposed by Kirkpatrick *et al.* [6] and Cerny *et al.* [7]. However some adaptations must be done to improve the algorithm's efficiency and effectiveness. These adaptations take into account the problem characteristics under study.

Simulated Annealing can be classified as a Local Search Meta-Heuristic. It can be initialized with a constructive heuristic and improved iteratively. For each iteration the algorithm selects a solution from the current neighborhood. In order to prevent the algorithm early stop on a local optimum, a mechanism based on the Metropolis Algorithm was incorporated, as well as the tuning of some parameters was undertaken, to guarantee efficiency and effectiveness. The parameters tuning are performed in: initial temperature (T_0), cooling schedule (R) and stop criterion. Other features are tailored based, such as:

- Objective function
- Initial solution generation
- Neighborhood function

The algorithm's procedure characterization, denominated by MonoSA, is shown in Figure 1. The MonoSA first sub-procedure (I) defines the initial solution generation. The second (II) addresses the neighbor solution, followed by the third sub-procedure (III) which deals with the new solution acceptance. The fourth (IV) analyzes the neighbor solution efficiency and, finally, the stop criterion and the restart mechanism are controlled by sub-procedure V.

Sub-Procedure I: The initialization of the SA parameters and generation of the initial solution for the algorithm to start is developed.

The initialization is made by feeding the algorithm with all the data and parameters, such as: number of products (P), machines (M) and orders (O), setups and processing times.

The initialization solution is obtained through the following steps:

1. Select the first order;
2. Select the machine with the lowest completion time;
3. Assign the order to the machine selected in step 2;
4. Go to step 2 until all the machines are selected, otherwise go to step 5;
5. Generate the orders' completion times by considering the setup times;
6. Compute the tardiness of each order. If the order is on time the tardiness is 0, otherwise has a delay;
7. The objective function is computed;

This sub-procedure concludes by setting the initial solution as the algorithm current solution.

Sub-Procedure II: this sub-procedure explores the search space, using the initial solution and the temperature (T) is update, through the classical geometric cooling mechanism. The temperature is periodically reduced from a relatively high value to near zero using Equation (1), as cooling schedule scheme.

Each temperature level (T_k) is maintained through a number of iterations (*NIST*).

$$T_{k+1} = RT_k \tag{1}$$

A neighbor solution is defined through a current solution movement. In this work is developed two types of movements. One movement exchanges the orders between different machines, while the other exchanges consecutive orders on a machine's line-up.

The second sub-procedure concludes with a new solution (s'_i) characterization.

Sub-Procedure III: this sub-procedure evaluates the new solution quality and its acceptance is analyzed.

If the neighbor solution (s'_i) has a better objective function value $(f_1(s'_i))$ then the new solution is automatically accepted, otherwise is accepted according to the acceptance probability, P_{ac}, defined in Equation (2), from Kirkpatrick *et al.* (1983)

$$P_{ac} = \begin{cases} 1 & f_1(s'_i) > f_1(s_i) \\ e^{\frac{f_1(s'_i) - f_1(s_i)}{T1}} & otherwise \end{cases} \tag{2}$$

Escaping from local optimum by accepting lower quality solutions is critical in SA. This probability decreases with the temperature's decrement as the method progresses. The acceptance probability also depends on the difference between objective function values. High differences lead to lower probabilities at the same temperature level.

Sub-Procedure IV: this sub-procedure analyzes if the current solution is better than the best solution so far, replacing it by the current solution, if necessary.

Fifth Sub-Procedure V: The length of the search is determined by the temperature, which starts with positive value and decreases as the search goes along, until the search is finally frozen.

In this work two stop criteria are used, the temperature lower than 0.00001 and the objective function $f(s_i)$ reaches the zero value (zero tardiness). The algorithm stops when one of the criteria is verified.

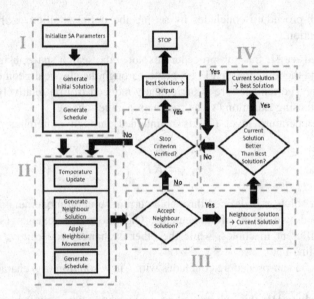

Fig. 1 MonoSA algorithm characterization.

3   Problem Characterization

Our case study focuses in a plastic containers facility. This type of industry operates under saturated markets with huge competitiveness levels and low unitary margins. The key factor for company's success is the final product price. Therefore cost minimization has a huge impact in its operations.

The facility under study receives weekly orders with different due dates to fulfil the customers demand, for the following week.

The production process is an injection molding characterized by several parallel machines and molds to produce containers as final products. Each container is characterized by two different parts, a bottom and a cover, with its own shape, label and color.

Every time a different containers order is put into production a set of changeover tasks, must be done. The changeover tasks must be performed every time a mold, dye or label must be changed, trigged by a new container characteristics. As an example, in some mold cases, several hours are required to conclude a change-over task. In this facility, the changeover time is not only product, but also machine complexity dependent and has a high impact in process productivity.

Nowadays, the facility operates under several limitations: for some products has high hand-on project inventory to overcome the demand uncertainty; a poor scheduling program neglecting products sequence and set-up time minimization, which is reflected in a high number of changeover tasks and customers' service level.

The aim of this work is to overcome the aforementioned limitation, enhancing not only the set-up minimization by reducing the number of changeover, but also developing an efficient production scheduling. The objective function most suitable for this propose is the Tardiness minimization.

4   Case Study

The aim of the proposed algorithm is to develop a scheduling of a plastic container facility to produce containers and fulfill weekly orders. Each container requires one bottom and one cover, and has its own characteristics, such as: shape, labels and colors. From now on a cover and a bottom are designated as products.

In the case under study 15 containers, equivalent to 30 products are produced, in 8 machines to fulfil 40 orders at the end of the week. An order is characterized by a product, a demand quantity and a due date. The processing and set-up time are machine and product dependent. For confidentiality reason, the process data are omitted.

To overcome some of the aforementioned limitation the algorithm performance is illustrated through two process strategies: the make-to-order (MTO) and the make-to-stock (MTS) strategy. The latter also considers inventory management.
In order to apply the algorithm, a parameters tuning were performed over the cooling rate (R), number of iteration at the same temperature level (NIST) and initial temperature (T_0). The cooling rate was analyzed over the values 0.95, 0.975 and 0.99, and the number iterations analyze over the values 100, 250 and 500. The initial temperature was selected empirically.

The best performance parameters values for both strategies were 1250 for temperature, and 0.99 for the cooling rate using 250 iterations.

4.1   Computational Results

In this paper the authors apply two strategies MTO and MTS to analyze the proposed algorithm performance using a Pentium 2 Duo, T7300, 2 GHz, 2 GB RAM.

The algorithm performance is evaluated for the same set of orders and its results are shown in Figure 2 and 3, for MTO and MTS, respectively.

The objective function values show different behavior between the strategies. The algorithm applied to a MTO strategy initiates with a higher objective function values and requires more iteration to reach the zero tardiness, when compared with the MTS. The MTS algorithm reaches the zero tardiness at an early iteration for all the tested parameters values, 100, 250 and 500 for the number of iterations and the values 0.925, 0.975, 0.99 for the cooling rate.

The higher performance from MTS strategy is justified by considering simultaneously the process schedule and inventory management in the algorithm. The orders considered for production follows the first-in-first-out rule, after inventory availability analysis. If the facility has enough on-hand inventory, the order is totally satisfied right away, otherwise the current need are quantified and its consolidation is analyzed before a production task is scheduled. If the due date allows orders consolidation, not only the set-up but also the tardiness minimized is performed.

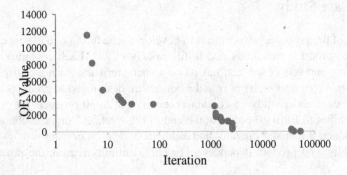

Fig. 2 MTO objective function results evolution.

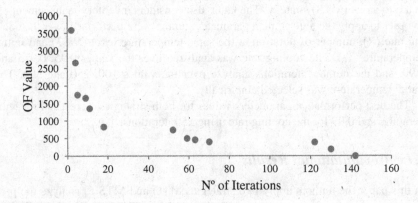

Fig. 3 MTS objective function results evolution.

The objective function values distribution over time was also analyzed in both strategies, for 200 runs, with a cooling rate of 0.99 and 500 iterations at the same temperature level. As is shown in Figure 3, the MTO strategy reached the optimal value in 78% of the runs. The remaining runs shows tardiness values higher than a working day shift, suggesting that is necessary additional tuning. The average CPU time for a run was 6.55 seconds.

For lack of space, the objective function value distribution over time for strategy MTS is omitted. However in the 200 runs the objective function converged to a zero tardiness in an average of 0.2 seconds, except one run with a tardiness of 100 min.

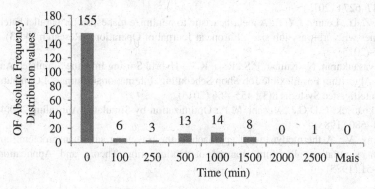

Fig. 4 MTO objective function distribution values.

5   Conclusions

In this paper the authors proposed a Simulating Annealing algorithm, MonoSA, with the total tardiness minimization as objective function, for the schedule of a container plastic injection molding process.

To illustrate the algorithm performance two strategies are compared, MTO and MTS. The MTO approach presented higher complexity, requiring higher temperatures and slower convergence compared with the MTS strategy. The MTS strategy reached zero tardiness in lower iteration.

The results show that process complexity has high impact in the algorithm performances and the MTS presents higher performance compared with the MTO strategy.

On-going research addresses the further tuning of the proposed SA approach and its application to more complex examples, in order to more firmly establish its potential as an alternative optimization approach to MILPs, when the application of the latter may lead to problem intractability.

Acknowledgments. This work was supported by Fundação para a Ciência e Tecnologia and Centro de Matemática e Aplicações through the project UID/MAT/00297/2013.

References

1. Chibeles-Martins, N., et al.: A meta-heuristics approach for the design and scheduling of multipurpose batch plants. In: 20th European Symposium on Computer Aided Process Engineering, vol. 28, pp. 1315–1320 (2010)
2. Chibeles-Martins, N., et al.: A simulated annealing approach for the biobjective design and scheduling of multipurpose batch plants. In: 21st European Symposium on Computer Aided Process Engineering, vol. 29, pp. 865–869 (2011)
3. Roshanaei, V., Azab, A., ElMaraghy, H.: Mathematical modelling and a meta-heuristic for flexible job shop scheduling. International Journal of Production Research 51(20), 6247–6274 (2013)
4. Jia, Z.-H., Leung, J.Y.T.: A meta-heuristic to minimize makespan for parallel batch machines with arbitrary job sizes. European Journal of Operational Research 240(3), 649–665 (2015)
5. Shivasankaran, N., Kumar, P.S., Raja, K.V.: Hybrid Sorting Immune Simulated Annealing Algorithm For Flexible Job Shop Scheduling. International Journal of Computational Intelligence Systems 8(3), 455–466 (2015)
6. Kirkpatrick, C.D.G.J., Vecchi, M.P.: Optimization by Simulated Annealing 220(4598), 671–680 (1983)
7. Cerny, V.: A thermodynamical approach to the travelling salesman problem: an efficient simulation algorithm. Journal of Optimization Theory and Applications 45, 41–51 (1985)

Applying Superquantile Regression to a Real-World Problem: Submariners Effort Index Analysis

Sofia Isabel Miranda

Marinha Portuguesa, Lisboa, Portugal
smiranda.pt@gmail.com

Abstract. We lay out the fundamental theory for superquantile regression. Such novel regression framework is centered on a coherent and averse measure of risk, the superquantile (also called conditional value-at-risk), which yields more conservatively fitted curves than classical least squares and quantile regressions. We illustrate this regression technique by analyzing a real-world problem where a random variable represents the effort index of the Portuguese Navy submariners along their Navy careers. This index was created as a decision tool to support human resource management inside the Submarine Squadron.

1 Introduction

Analyzing real-world problems is a challenge to decision makers, and one of such problems occurs when one is faced with a random variable describing possible outcomes of a stochastic phenomenon. One seeks to quantify the uncertainty in such a loss random variable[1] by surrogate estimation. Since its upper-tail realizations are considered prejudicial, underestimating is considered more detrimental than overestimating them. As relevant statistics of this loss random variable are rarely available, a possibility is to attempt to approximate such loss random variable by an explanatory random vector that is more accessible.

In this paper[2], we lay out the superquantile regression problem, and present a computational method for solving this problem based on the dualization of risk. We finally illustrate this regression technique by analyzing the effort index of Portuguese Navy submariners.

---

[1] We denote this by "loss random variable" although it could describe possible "cost".

[2] The fundamental theory presented in this paper is to a large extend based on Rockafellar et al. (2014), and Miranda (2014). Proofs can be seen in more detail in Miranda (2014).

A.P.F.D. Barbosa Póvoa and J.L. de Miranda (eds.), *Operations Research and Big Data*,
Studies in Big Data 15, DOI: 10.1007/978-3-319-24154-8_14

2   Superquantile Regression Problem

We consider a loss random variable Y as a function on a probability space $(\Omega, \mathcal{F}, P)$, and we assume that Y has a finite second moment. Ω is a sample space with $\omega \in \Omega$ being a possible outcome; $\mathcal{F}$ is an event space; and P is a probability measure that assigns probabilities to these events, $P : \mathcal{F} \to [0, 1]$,

$$Y \in \mathcal{L}^2 := \mathcal{L}^2(\Omega, \mathcal{F}, P) := \{Y : \Omega \to I\!R \mid Y \ \mathcal{F}\text{-measurable}, E[Y^2] < \infty\}.$$

Since the distributional information of Y is rarely available, we approximate Y by an n-dimensional explanatory random vector X that is more accessible, and for a carefully selected regression function $f : I\!R^n \to I\!R$, the random variable $f(X)$ serves as a surrogate of Y. Clearly, one would like the error random variable $Z_f := Y - f(X)$ to be small in some sense. Least squares regression is obtained by minimizing $E[Z_f^2]$, while quantile regression is obtained by miniming the Koenker-Bassett measure of error (Koenker, 2005), but many other possibilities exist.

The "Fundamental Risk Quadrangle" is a concept introduced by Rockafellar and Uryasev (2013), which establishes the connections between distinct measures of a random variable whose orientation is such that large realizations are unfortunate and low realizations are favorable. The interrelationships of such numerical quantities allow distinct comparisons or application in various analysis, such as risk management. We consider the following distinct functionals on $\mathcal{L}^2$ that assign numerical values to e.g., a loss random variable Y. A *measure of error* $\mathcal{E}(Y)$ quantifies the "nonzeroness" in Y, e.g., the $\mathcal{L}^2$-norm of Y. A *measure of risk* $\mathcal{R}(Y)$ serves as surrogate for the overall loss in Y, e.g., the essential supremum, $\mathcal{R}(Y) = \sup\{Y\}$, or less conservatively $\mathcal{R}(Y) = E[Y]$. A *measure of deviation* $\mathcal{D}(Y)$ quantifies the "nonconstancy" as uncertainty in Y, and can be seen as a generalization of the standard deviation of Y. A *measure of regret* $\mathcal{V}(Y)$ quantifies the displeasure of obtaining mix realizations of Y, which might be better when $Y \leq 0$ (representing "gains") or worse when $Y > 0$ (representing "losses"). And a *statistic* $\mathcal{S}(Y)$ is associated with Y through $\mathcal{E}$ and $\mathcal{V}$.

Diagram 3 in Rockafellar and Uryasev (2013), defines these distinct measures of Y and their general relationships, as listed below.

Error measure $= \mathcal{E}(Y) = \mathcal{V}(Y) - E[Y]$
Risk measure $= \mathcal{R}(Y) = \min_{c_0}\{c_0 + \mathcal{V}(Y - c_0)\}$
Deviation measure $= \mathcal{D}(Y) = \mathcal{R}(Y) - E[Y]$
Regret measure $= \mathcal{V}(Y) = \mathcal{E}(Y) + E[Y]$
Statistic $= \mathcal{S}(Y) = \arg\min_{c_0}\{c_0 + \mathcal{V}(Y - c_0)\} = \arg\min_{c_0}\{\mathcal{E}(Y - c_0)\}$

We focus on regression functions of the form

$$f(x) = c_0 + \langle c, h(x) \rangle, \qquad c_0 \in I\!\!R, c \in I\!\!R^m,$$

for a given "basis" function $h : I\!\!R^n \to I\!\!R^m$. This class satisfies most practical needs including that of linear regression where $m = n$ and $h(x) = x$. From Rockafellar et al. (2014), the *Superquantile Regression Problem P*, for any $h : I\!\!R^n \to I\!\!R^m$ and $\alpha \in (0,1)$, is defined as

$$P : \min_{c_0 \in I\!\!R, c \in I\!\!R^m} \bar{\mathcal{E}}_\alpha (Z(c_0, c)) = \frac{1}{1-\alpha} \int_0^1 \max\{0, \bar{q}_\beta(Z(c_0, c))\} d\beta - E[Z(c_0, c)],$$

where $Z(c_0, c) := Y - (c_0 + \langle c, h(X) \rangle)$ is the error random variable, whose distribution depends on c_0, c, h, and the joint distribution of (X, Y). We denote by $\bar{C} \subset I\!\!R^{m+1}$ the set of optimal solutions of P and refer to $(\bar{c}_0, \bar{c}) \in \bar{C}$ as a regression vector.

As a direct consequence of the Regression Theorem in Rockafellar and Uryasev (2013) (see also Theorem 3.1 in Rockafellar et al., 2008), we obtain that a regression vector can equivalently be determined from a measure of deviation $\bar{\mathcal{D}}_\alpha$. According to Proposition II.3 in Miranda (2014), we obtain the following equivalent problem, the *Deviation-based Superquantile Regression Problem D*, for any $h : I\!\!R^n \to I\!\!R^m$ and $\alpha \in (0,1)$,

$$D : \min_{c \in I\!\!R^m} \bar{\mathcal{D}}_\alpha (Z_0(c)) = \frac{1}{1-\alpha} \int_\alpha^1 \bar{q}_\beta(Z_0(c)) d\beta - E[Z_0(c)],$$

with $\bar{c}_0$ being obtained by setting $\bar{c}_0 = \bar{q}_\alpha(Z_0(\bar{c}))$. This implies computational advantages as the $(m+1)$-dimensional optimization problem P is replaced by a problem in m dimensions with a simpler objective function

We refer to Rockafellar et al. (2014), and Miranda (2014) for existence and uniqueness, consistency and stability of the regression vector, and rate of convergence results.

3 Computational Methods

We present a class of computational methods that allow us to solve a convex optimization problem: the superquantile regression problem P. Due to incomplete distributional information, in practice we seek to solve an approximate problem P^ν, where ν represents the number of observations. Regardless of the distribution of (X^ν, Y^ν), a reformulation of the approximate problem P^ν in terms of the deviation measure $\bar{\mathcal{D}}_\alpha$ is beneficial.

As described in detail in Miranda (2014), the deviation measure for our superquantile-based quadrangle is defined as follows

$$\bar{\mathcal{D}}_\alpha(Y) = \bar{\mathcal{R}}_\alpha(Y) - E[Y] = \frac{1}{1-\alpha} \int_\alpha^1 \bar{q}_\beta(Y) d\beta - E[Y] = \bar{\bar{q}}_\alpha(Y) - E[Y], \quad (1)$$

where $\bar{\mathcal{R}}_\alpha(Y) = \bar{\bar{q}}_\alpha(Y)$ is the α-second-order superquantile risk measure for which we build the dual for.

By the Envelope Theorem in Rockafellar and Uryasev (2013), an alternative formula for a positively homogeneous regular risk measure[3] $\mathcal{R}(\cdot)$ is given by its dual representation, described as follows

$$\mathcal{R}(Y) = \sup_{Q \in \mathcal{Q}} \{E[YQ]\}, \tag{2}$$

where $\mathcal{Q}$ is a nonempty closed convex set that is to the risk envelope associated with $\mathcal{R}$. For $Y \in \mathcal{L}^2$ and $\alpha \in (0,1)$, a Q^Y that maximizes

$$\sup_{Q \in \mathcal{Q}} \{E[YQ]\}$$

is called a risk identifier. And we are able to define the objective function of our new problem as follows

$$f(c) = \frac{1}{\nu} \sum_{i=1}^{\nu} Z_0(c)^{(i)} \bar{Q}_\alpha^{Z_0(c)}(i) - \frac{1}{\nu} \sum_{j=1}^{\nu} Z_0(c)^j \tag{3}$$

$$= \frac{1}{\nu} \sum_{i=1}^{\nu} \left(y^{(i)} - \langle c, h(x^{(i)}) \rangle \right) \bar{Q}_\alpha^{Z_0(c)}(i) - \frac{1}{\nu} \sum_{j=1}^{\nu} \left(y^j - \langle c, h(x^j) \rangle \right),$$

where $Z_0(c)^{(i)}$ is the ith-ordered value of $Z_0(c)$. The evaluation of the objective function requires the computation of $\bar{Q}_\alpha^{Z_0(c)}$. According to Proposition III.3 in Miranda (2014), this implies sorting vector $Z_0(c)$ for a given c to obtain its cumulative distribution function and only then evaluate $\bar{Q}_\alpha^{Z_0(c)}$, using the same sorting as for $Z_0(c)^{(i)}$. A subgradient of $f(c)$ is then easily computed as follows

$$\nabla f(c) = -\frac{1}{\nu} \sum_{i=1}^{\nu} h\left(x^{(i)}\right) \bar{Q}_\alpha^{Z_0(c)}(i) + \frac{1}{\nu} \sum_{j=1}^{\nu} h\left(x^j\right), \tag{4}$$

with $h\left(x^{(i)}\right)$ maintaining the same ordering as in $Z_0(c)^{(i)}$ used in (3), and allowing us to implement well-known algorithms, e.g., cutting plane method.

We next apply these theoretical results using the Portuguese Navy submariners effort index as our real-world problem.

4   Portuguese Navy Submariners Effort Index

We consider a data set with 103 observations provided by the Portuguese Navy Submarine Squadron and seek to estimate a random variable Y representing the effort index of the Portuguese submariners along their Navy

---

[3] A *regular measure of risk* satisfies: $\mathcal{R}(c) = c$ for a constant c, as well as convexity, aversion, and closedness, $\{Y \mid \mathcal{R}(Y) \leq c\}$ for all constants $c \in I\!R$.

careers[4]. This index was created as a decision tool to support human resource management inside the Submarine Squadron. Once a sailor becomes a submariner, his career depends mainly on the Submarine Squadron. The Commanding Officer of the Submarine Squadron has the power of assigning a submariner that is serving at the Submarine Squadron for a mission, if there is the need to embark an extra element or substitute someone onboard. It is crucial to support such decision with a tool that emphasizes who is more "available" for the mission.

We assume higher effort indices to be more prejudicial than small indices for the accomplishment of the Submarine Squadron mission, i.e., overemploying is worse than underemploying a submariner with consequences that we do not enumerate here. The idea behind this index is to better analyze submariners careers which helps determine selection criteria for future Submarine Squadron personnel recruitment and also understand who has been overemployed. Therefore, errors are seen asymmetrically and most of all their magnitudes are critical.

One of the goals with this example is to show that superquantile regression helps us better visualize what may cause the discrepancies in effort indices among submariners and take the obtained results to help manage human resources.

We consider five possible explanatory variables: years since a submariner has gained the insignia of the Portuguese submarine service (X_{dolphins}), years a submariner has embarked on surface warships (X_{surf}), years a submariner has been ashore (X_{ashore}), total submarine navigation hours (X_{sub}), and submariners age (X_{age}). Since there is a strong linear correlation between the effort index Y and the explanatory variable X_{sub}, and since X_{surf} and X_{ashore} do not play a decisive role if added to our model, we only explore linear and quadratic models using both variables X_{dolphins} and X_{age} as factors. Naturally, one thinks that the larger X_{dolphins}, the higher the effort index. Figure 1 shows how the effort index behaves with X_{dolphins}. One expects similar results with X_{age}. The idea of older submariners having more experience due to more training has not always been true, and that issue raised the question of how to quantify training and expertise. For example, a 39-year old submariner can have an effort index as low as 5 or as high as 22, as seen in Figure 2. Such discrepancies cause discomfort among fellow submariners.

A small detail that we encounter in Figure 1, is the lack of observations for values of X_{dolphins} between 4 and 7 years, which is due to the fact that Portugal acquired the Tridente-class submarines in 2010, and the few years prior were dedicated to training the existing submariners to a completely new technology. This process required the Portuguese Navy to delay the submariners course until after the reception of the new assets.

---

[4] Computations are carried out in Matlab version 7.14 on a 2.26 GHz laptop with 8.0 GB of RAM using Portfolio Safeguard (see American Optimal Decisions, Inc., 2011), with VAN as the optimization solver for the superquantile regression problem.

First, we start with a linear model f_1 with only one explanatory variable at a time, X_{dolphins} or X_{age}. Then we consider a linear model f_2 with both explanatory variables, X_{dolphins} and X_{age}. Table 1 presents the obtained solution vectors and the corresponding coefficients of determination[5], for a probability level $\alpha = 0.75$. Rows 4, and 8 report the regression vectors for model f_2. We realize that the coefficient of determination improves in the cases of f_2.

Table 1 Regression vectors and coefficients of determination $\bar{R}^2_{0.75}$ for linear and quadratic models at a fixed probability level $\alpha = 0.75$.

Regression	Model	c_0	c_{dolphins}	c_{age}	c_{age2}	$\bar{R}^2_{0.75}$
Least Squares	f_1	3.1365	0.8643	—	—	0.7452
		-17.6369	—	0.7983	—	0.4845
	f_2	8.1218	0.9918	-0.1711	—	0.7512
	f_3	-87.1182	—	4.6498	-0.05251	0.5442
Superquantile	f_1	2.9811	1.2172	—	—	0.5866
		-27.0234	—	1.2048	—	0.2403
	f_2	7.3697	1.3430	-0.1558	—	0.5939
	f_3	-126.4859	—	6.9812	-0.0827	0.3235

In Figure 1, we report the linear model $f_1(x) = c_0 + c_{\text{dolphins}}x_{\text{dolphins}}$ for least squares, 0.60-quantile and 0.60-superquantile regressions. All three obtained regression functions have completely distinct slopes. The dotted line representing the 0.60-quantile regression fit gives us the notion of where the 40% worst cases are, while the dashed line representing the 0.60-superquantile regression function provides the average of the 40% worst effort indices, characterized by those observations above the 0.60-quantile regression function.

Second, we consider a quadratic model of the form $f_3(x) = c_0 + c_{\text{age}}x_{\text{age}} + c_{\text{age2}}x^2_{\text{age}}$. Rows 5, and 9 of Table 1 show the obtained regression vectors and the corresponding coefficients of determination. The obtained coefficients of determination for the quadratic model are bigger than those for the linear models that only include the explanatory variable X_{age}, and we plot the regression functions in Figure 2. Here the 0.75-superquantile regression function captures the effects of the higher effort indices and forms an interesting curvature. The 0.75-quantile regression model on the other hand is not affected by these observations and it looks almost parallel to the least squares regression model for 40-year old submariners.

Here we have a real-world example where magnitudes are as important as the signs of the regression errors. Obviously, we would need to perform further analysis to better understand the submariners effort index and obtain

---

[5] The coefficient of determination and other goodness of fit analysis tools applied to superquantile regression are defined in Miranda (2014).

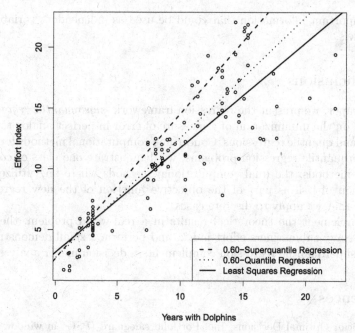

Fig. 1 Regression lines for model $f_1(x) = c_0 + c_{\text{dolphins}}x_{\text{dolphins}}$, at probability level $\alpha = 0.60$.

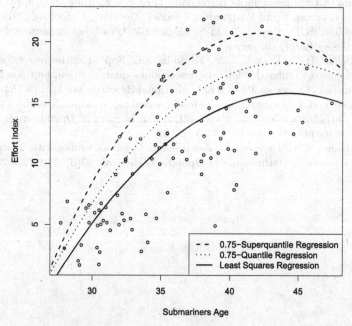

Fig. 2 Quadratic regression models $f_3(x) = c_0 + c_{\text{age}}x_{\text{age}} + c_{\text{age2}}x_{\text{age}}^2$, at probability level $\alpha = 0.75$.

other important information that could be used as independent variables in
our models.

5 Conclusions

In this paper, we present the regression framework, *superquantile regression*,
centered on the minimization of a measure of error in perfect analog to least
squares and quantile regressions. Concerning computational methods for solving superquantile regression problems, we demonstrate one class of computational methods: the dual computational methods where one utilizes the
dualization of risk as part of the objective function of the new regression
problem that we apply to discrete cases.

We implement the theoretical results in a real-world problem, the Portuguese Navy submariners effort index, and perform a small demonstration
on how superquantile regression complements a decision maker analysis.

References

1. American Optimal Decisions, Inc. Portfolio safeguard (PSG) in windows shell
   environment: Basic principles. AORDA, Gainesville, Florida (2011)
2. Koenker, R.: Quantile regression. Cambridge University Press, Cambridge (2005)
3. Miranda, S.I.: Superquantile regression: Theory, algorithms, and applications.
   Phd dissertation, Naval Postgraduate School, Monterey, California (2014)
4. Rockafellar, R.T., Royset, J.O.: Superquantile/CVaR risk measures: Second-
   order theory (2014) (in review)
5. Rockafellar, R.T., Royset, J.O., Miranda, S.I.: Superquantile regression with
   applications to buffered reliability, uncertainty quantification, and conditional
   value-at-risk. European Journal of Operational Research 234(1), 140–154 (2014)
6. Rockafellar, R.T., Uryasev, S.: The fundamental risk quadrangle in risk management, optimization and statistical estimation. Surveys in Operations Research
   and Management Science 18, 33–53 (2013)
7. Rockafellar, R.T., Uryasev, S., Zabarankin, M.: Risk tuning with generalized
   linear regression. Mathematics of Operations Research 33(3), 712–729 (2008)

Recent Trends and Challenges in Planning and Scheduling of Chemical-Pharmaceutical Plants

Samuel Moniz[1], Ana Paula Barbosa-Póvoa[2], and Jorge Pinho de Sousa[1,3]

[1] INESC TEC, Rua Dr. Roberto Frias, Porto, Portugal
[2] Centro de Estudos de Gestão, Instituto Superior Técnico, Universidade de Lisboa,
  Av. Rovisco Pais, Lisboa, Portugal
[3] Faculty of Engineering, University of Porto, Rua Dr. Roberto Frias, Porto, Portugal

Abstract. This paper discusses the current trends in optimization methods for solving planning and scheduling problems in the chemical-pharmaceutical industry. The challenges of this industry and the recent advances in modeling these problems show that optimization methods need to provide highly integrated solutions encompassing decision-making at both R&D and Operations levels. The heterogeneous demand, characteristic of the complex drug development cycle, asks for mixed planning strategies capable of increasing the resources utilization and the plant output, and of dealing with uncertainty.

1 Introduction

Industrial companies are continuously assessing their operations, as a way to increase the overall effectiveness of production systems. Markets where these organizations operate tend to become more complex over time, forcing companies to increase their responsiveness, both in terms of time and cost. The case of the pharmaceutical industry is a good example on how market is driving the change of the drug development cycle and manufacturing activities. Some of the most relevant driving factors are related to: a) the drought in new drug approval applications by the regulatory agencies; b) the uncertainty associated to the Research and Development (R&D) activities and trials I-III phases; and c) the pressure on the drug prices and demand variability, caused by patent drops. This context is putting enormous pressure in the industry to reduce the time and the cost required to launch new drugs to market and, when drugs are being commercialized, to reduce the manufacturing and inventory costs and the typically long production lead times [1]. The planning and scheduling functions are then critical in these highly dynamic production systems.

The relevance of using optimization tools to solve planning and scheduling problems has been recognized by the industry [2]. Thus, not surprisingly, the efforts made in the past years, by academia and industry, resulted into several successful cases of integration of optimization tools into complex decision-making

© Springer International Publishing Switzerland 2015

A.P.F.D. Barbosa Póvoa and J.L. de Miranda (eds.), *Operations Research and Big Data,*
Studies in Big Data 15, DOI: 10.1007/978-3-319-24154-8_15

processes related to supply chain, process design, planning and scheduling [3]. However, despite the significant progress achieved in modeling these problems, the development of efficient optimization and solution methods, capable to be integrated in corporate decision-making, is still an open research topic. This paper describes the major issues in planning and scheduling decision-making in the chemical-pharmaceutical industry, and reviews some important optimization methods that have been applied in this context. The remainder of the paper is structured as follows. Section 2 presents some recent trends in modeling planning and scheduling problems in the case of the chemical-pharmaceutical industry. The characteristics of this industry are described so as to provide a common understanding of the several complexities associated with planning and scheduling problems. Section 3 summarizes the challenges and opportunities in this field. Finally, in section 4, some concluding remarks are presented.

2   Planning and Scheduling Decision-Making

Manufacturers and regulators in the pharmaceutical industry create a specific context for Operations Management [4], thus conditioning planning and scheduling functions. The planning problem involves the determination of tactical production plans, in which decisions are typically made assuming a certain degree of aggregation of resources and time, hence imposing bounds to the scheduling problem. The scheduling problem involves the determination of operational plans at the level of the most elementary production resources and at a fine time grid. In practice both problems may present several conflicting objectives such as, for example, minimizing costs and minimizing delivery times. Moreover the characteristics of the market, of production processes, and of chemical plants make planning and scheduling tasks particularly difficult to perform [5].

The drug development cycle imposes the coordination of a large number of different R&D and manufacturing activities, and therefore planning and scheduling decision-making is essential to achieve a global optimization of these systems. In this context, solving planning and scheduling problems requires the application of different methods that are briefly presented next (due to space limitations, we do not to provide here an extensive review on this topic):

Planning and Scheduling of Products under Development

The planning and scheduling functions are deeply coupled with R&D tasks (Fig. 1). Process Design consists in using the information available to develop an industrial process. Here, the characteristics of the products and processing units are considered in developing an efficient production process concerning the resources utilization, given a set of market and operating constraints. And Production Execution and Control involve production dispatching, control and quality assessment, among other activities.

Planning and scheduling encompass then the coordination of development and manufacturing activities, so as to move from the laboratory scale to the industrial scale, this resulting in the determination of the final product quantities (lot sizes) and in a first assessment of the processing times. The goal is to find optimal schedules that maximize the expected economic value of the investment by considering the resources availability, the probability of success of the clinical trials, and the associated costs. Varma et al. [6] have developed a comprehensive decision-making framework called Sim-Opt for resource management, that includes components for stochastic simulation, schedules generation based on a mixed integer linear programming (MILP) formulation, and evaluation of various resource strategies. A common modeling method for solving the clinical trials planning problem is multi-stage stochastic programing, coupled with solution methods, such as specialized branch and cut algorithms, Lagrangean decomposition, Benders decomposition, as a way to improve the computational performance [7,8].

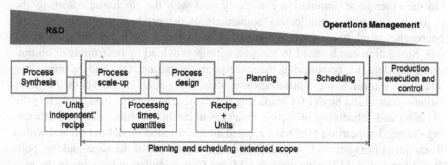

Fig. 1 Scope of the planning and scheduling problems. Source [5]

Integration of Planning and Scheduling

It is common to find chemical plants that manufacture products with dissimilar production recipes and volumes. For low volume production (short-term mode) a reduced number of batches are produced and for high volume production (campaign mode) the number and size of batches tends to be higher. These strategies have impact on how resources are allocated and on the system responsiveness. The short-term mode, under a multipurpose environment, requires a higher responsiveness from the production system, since resources are shared between several products in a very dynamic production environment, whereas in the campaign mode, resources are allocated to single products during long time periods. Campaign schedules can be computed using a periodic scheduling approach [9], leading to schedules that are seen as operationally easier to manage and execute. The combination of both planning modes (short/campaign) is often present when a mixed planning strategy is explored. In these cases short-term demand can often be planned to follow a non-periodic scheduling, and campaigns are generated to follow a periodic scheduling [10]. Alternatively, planning and scheduling problems have been simultaneously solved using multiscale methods. These methods account for the impact of the mid-term decisions in the short-term decisions (and vice-versa) and integrate the decision-making processes of different planning levels [11,12]. Multiscale methods propose decomposition schemes that include models with different planning horizons and different aggregation levels for the associated problems, and aim at increasing the resources utilization and the facility output [13]. So relevant data for decision-making is modeled with different time scales depending on its availability and reliability.

Finally, a reference should be made to the integration of planning and scheduling with real time optimization/control, in order to improve the performance of the production systems at all hierarchical decision-making levels [14].

Production Simulation-Optimization

Modeling the uncertainty of real world problems with optimization often results into models with a large number of variables, that are computationally intractable. On the other hand, simulation can easily deal with the stochastic nature of the problems and with complicating constraints of the production systems. Recent approaches combine simulation and optimization in new, efficient solution methods. Simulation can be used to assess solutions provided by deterministic optimization models, by considering the uncertainty of parameters, and by providing additional information to optimization models. Chen et al. [15] developed a simulation-optimization model for managing the entire trial supply chain, including the planning and scheduling of Active Pharmaceutical Ingredients (API) manufacturing. Sahay, Ierapetritou [16] have applied an agent-based model hybridized with a linear programing model, to study different supply chain decision-making policies. Eberle et al. [17] have applied a Monte-Carlo simulation to estimate the production lead time of pharmaceutical processes.

Ontologies and Knowledge Management

Models to support planning and scheduling decisions are part of complex software environments that tend to share data structures and that involve many information flows. Zhao et al. [18] introduced a web-based infrastructure called *Pharmaceutical Informatics* that aims at supporting the overall decision-making process and includes the following areas: product portfolio, capacity planning, planning and scheduling, process control and safety, and supply chain management. Muñoz et al. [19] proposed an ontological framework for streamlining the exchange of information and knowledge models between different applications and hierarchical decision levels. The framework has been demonstrated in a supply chain network design-planning problem. Focusing on solving the scheduling problem, Moniz et al. [20] proposed a methodology that integrates the representation of the scheduling problem, the optimization model, and the decision-making process. The input data required for the MILP they have developed is automatically captured by a novel process representation tool used by several departments of the company.

3 Challenges and Opportunities

Despite the significant academic and industrial achievements in this area, there are relevant challenges that make planning and scheduling decision-making particularly difficult to address. Solving the integrated problem requires the development of comprehensive optimization tools that should be integrated with upstream design decisions and with downstream process control, in order to deal with the complex tradeoffs between development and manufacturing. In fact, understanding and modeling the conditions for achieving the global optimization of these manufacturing systems is by itself a quite hard task. To address this issue, we have proposed in a previous work, a representation of R&D and manufacturing tradeoffs, called Delivery Trade-offs Matrix (DTM) (**Fig. 2 a**) [5]. The DTM shows the three phases of the drug development cycle (R&D, trials I-III, and commercialization) against some relevant issues this industry needs to continuously manage, such as cost, uncertainty, time-to-market, amount delivered, campaign and short-term planning. The size of the "bubble" at each phase is proportional to the lot size, showing that a batch plant may need to produce from few grams to hundred kilos of the same product.

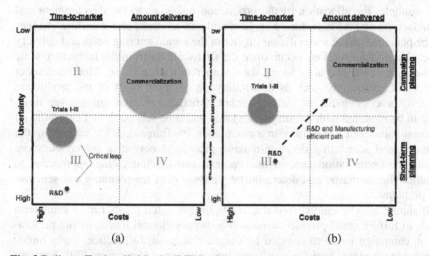

Fig. 2 Delivery Trade-offs Matrix (DTM) of the pharmaceutical industry: a) current state, b) future scenario. Source [5]

Several different process scale-ups are typically performed to respond to the demands at the early stages of the development, at the clinical trials I-III, and also after the drug approval for commercialization. Therefore the development of methods capable of addressing the long–term dimension of the scale-up decisions is surely a promising topic of research. The delivery of drugs for Trials I-III is of extreme importance, due to the fact that the clinical trials must be rigorously followed, and also because it is common to have more than one company developing similar drugs, time-to-market being therefore essential (Fig. 2 a). On the other hand, after launching the drug in the market there is more flexibility concerning delivery dates, since typically there are more inventory on the supply chain [21]. So, while in the R&D and trials I-III phases, time-to-market is the main driver, in the commercialization phase delivering the right product amounts appears to be more relevant.

In this industry, probably more than in other sectors, the product is strongly linked to the process. The way chemical processes are designed, implemented, controlled, and scaled-up strongly determines the overall cost and quality of the product, the total production time, and the global efficiency of the multipurpose batch plant. So, it is interesting to extend the scope of the planning and scheduling functions to account for design problems [22]. This is especially useful when manufacturing involves chemical processes that are under development. The solution space of planning and scheduling decisions will benefit with the integration of scale-up and process design decisions, and will potentially lead to a better utilization of the available resources, thus increasing the plant output. In general, the integration of models for the different decision levels still needs to be improved, to avoid suboptimal solutions and to better represent the real information flow of the decision making process.

Uncertainty and costs are quite complex to deal with, since the knowledge at the R&D phase is seldom sufficient to ensure a successful scale up to the plant [1]. For example, the allocation of the production resources to the development and manufacturing of APIs is done 6 to 12 months in advance, and therefore changes in the plans may have a significant impact in the manufacturing costs and delivery time. Demand fluctuations occur since the first scale up from the laboratory scale to the plant scale, and also during the commercialization phase. Moreover, other sources of uncertainty such as uncertainty on complex steps of the production process, processing times, yields, machine breakdowns, and changeover times need to be modeled with the different spatial and time scales. In fact capturing all relevant sources of uncertainty in a comprehensive framework for supporting the planning and scheduling decision-making process is not yet a solved question. Simulation-optimization can, in this context, provide interesting approaches to combine the stochastic and deterministic parameters of the planning and scheduling problems.

It should also be emphasized that multipurpose batch plants run in short-term mode to fulfil a small product demand (*e.g.* supply clinical trials), or run preferably in campaign mode to respond to a regular demand. In practice, many multipurpose batch plants simultaneously produce both types of products [20], however a reduced number of research works propose solution methods capable of addressing mixed planning strategies.

The complexity and variety of the decisions that need to be made during a drug development cycle require highly integrated information flows. We can therefore expect that the development of ontologies for building comprehensive frameworks will foster the integration of model-based applications by this industry. The chemical-pharmaceutical industry has to systematically address the huge uncertainty present in the supply chain in order to decrease the development and manufacturing time and cost (Fig. 2 b). The path to efficient R&D and manufacturing activities requires integrated decision-making and multiscale modeling so as to account with a rather heterogeneous demand.

4 Conclusions

This paper discusses some current trends in optimization methods for solving planning and scheduling problems in the chemical-pharmaceutical industry. Recent research works address integration, uncertainty, knowledge management and modeling issues, to solve extended planning and scheduling problems. Some of these works propose methods for solving real world problems and show the benefits of the deployment of model-based applications in the industry. In fact the need for innovative, integrated solution approaches is also justified by the pressure to reduce R&D and manufacturing time and costs but, despite the significant academic and industrial achievements in this area, there are still quite relevant challenges to address.

Acknowledgments. This research was partially financed by project NORTE-07-0124-FEDER-000057, the North Portugal Regional Operational Programme (ON.2 - O Novo Norte), under the National Strategic Reference Framework (NSRF), through the European Regional Development Fund (ERDF), and by national funds, through the Portuguese funding agency, Fundação para a Ciência e a Tecnologia (FCT).

References

1. Federsel, H.-J.: Chemical Process Research and Development in the 21st Century: Challenges, Strategies, and Solutions from a Pharmaceutical Industry Perspective. Accounts of Chemical Research 42(5), 671–680 (2009). doi:10.1021/ar800257v
2. Klatt, K.-U., Marquardt, W.: Perspectives for process systems engineering—Personal views from academia and industry. Comput. Chem. Eng. 33(3), 536–550 (2009)
3. Grossmann, I.: Enterprise-wide optimization: A new frontier in process systems engineering. AIChE Journal 51(7), 1846–1857 (2005)
4. Laínez, J.M., Schaefer, E., Reklaitis, G.V.: Challenges and opportunities in enterprise-wide optimization in the pharmaceutical industry. Comput. Chem. Eng. 47, 19–28 (2012)
5. Moniz, S., Barbosa Póvoa, A.P., Pinho de Sousa, J.: On the complexity of production planning and scheduling in the pharmaceutical industry: the Delivery Trade-offs Matrix. Computer Aided Chemical Engineering 37, 1865–1870 (2015)
6. Varma, V.A., Pekny, J.F., Blau, G.E., Reklaitis, G.V.: A framework for addressing stochastic and combinatorial aspects of scheduling and resource allocation in pharmaceutical R&D pipelines. Comput. Chem. Eng. 32(4), 1000–1015 (2008)
7. Colvin, M., Maravelias, C.T.: Modeling methods and a branch and cut algorithm for pharmaceutical clinical trial planning using stochastic programming. European Journal of Operational Research 203(1), 205–215 (2010)
8. Sundaramoorthy, A., Li, X., Evans, J.M., Barton, P.I.: Capacity planning under clinical trials uncertainty in continuous pharmaceutical manufacturing, 2: solution method. Industrial & Engineering Chemistry Research 51(42), 13703–13711 (2012)

9. Shah, N., Pantelides, C., Sargent, R.: Optimal periodic scheduling of multipurpose batch plants. Annals of Operations Research 42(1), 193–228 (1993)

10. Moniz, S., Barbosa-Póvoa, A.P., Pinho de Sousa, J.: Simultaneous regular and non-regular production scheduling of multipurpose batch plants: A real chemical–pharmaceutical case study. Comput. Chem. Eng. 67, 83–102 (2014)

11. Stefansson, H., Shah, N., Jensson, P.: Multiscale planning and scheduling in the secondary pharmaceutical industry. AIChE Journal 52(12), 4133–4149 (2006)

12. Amaro, A., Barbosa-Póvoa, A.: Planning and scheduling of industrial supply chains with reverse flows: A real pharmaceutical case study. Comput. Chem. Eng. 32(11), 2606–2625 (2008)

13. Maravelias, C.T., Sung, C.: Integration of production planning and scheduling: Overview, challenges and opportunities. Comput. Chem. Eng. 33(12), 1919–1930 (2009)

14. Chu, Y., You, F.: Model-based integration of control and operations: Overview, challenges, advances, and opportunities. Comput. Chem. Eng. (2015)

15. Chen, Y., Mockus, L., Orcun, S., Reklaitis, G.V.: Simulation-optimization approach to clinical trial supply chain management with demand scenario forecast. Comput. Chem. Eng. 40, 82–96 (2012)

16. Sahay, N., Ierapetritou, M.: Hybrid Simulation Based Optimization Framework for Centralized and Decentralized Supply Chains. Industrial & Engineering Chemistry Research 53(10), 3996–4007 (2014)

17. Eberle, L.G., Sugiyama, H., Schmidt, R.: Improving lead time of pharmaceutical production processes using Monte Carlo simulation. Comput. Chem. Eng. 68, 255–263 (2014)

18. Zhao, C., Joglekar, G., Jain, A., Venkatasubramanian, V., Reklaitis, G.: Pharmaceutical informatics: A novel paradigm for pharmaceutical product development and manufacture. Computer Aided Chemical Engineering 20, 1561–1566 (2005)

19. Muñoz, E., Capón-García, E., Laínez, J.M., Espuña, A., Puigjaner, L.: Integration of enterprise levels based on an ontological framework. Chemical Engineering Research and Design (2013)

20. Moniz, S., Barbosa-Póvoa, A.P., Pinho de Sousa, J., Duarte, P.: A solution methodology for scheduling problems in batch plants. Industrial & Engineering Chemistry Research (2014). doi:10.1021/ie403129y

21. Shah, N.: Pharmaceutical supply chains: key issues and strategies for optimisation. Comput. Chem. Eng. 28(6-7), 929–941 (2004)

22. Barbosa-Povoa, A.P.: A critical review on the design and retrofit of batch plants. Comput. Chem. Eng. 31(7), 833–855 (2007). doi:10.1016/j.compchemeng.2006.08.003

Planning Collection Routes with Multi-compartment Vehicles

Adriano Dinis Oliveira, Tânia Rodrigues Pereira Ramos and Ana Lúcia Martins

Instituto Universitário de Lisboa (ISCTE-IUL), Business Research Unit (BRU), Lisboa, Portugal

Abstract. This paper aims to assess the impact of using vehicles with multiple compartments in a recyclable waste collection system. Such systems perform single-material routes to collect three types of recyclable materials (paper, glass and plastic/metal), where vehicles with a single compartment are used. If vehicles with multi-compartments were used, two or even the three materials could be collected simultaneously without mixing them. A heuristic approach is developed to solve the multi-compartment vehicle routing problem and applied to a real waste collection system operating in Portugal.

1   Introduction

Material recycling, imposed by the European Union (EU), brought extra logistics challenges to the member states. Collection costs represent about 70% of the total costs for a recyclable waste collection system (Ramos et al. 2014). Given this figure, such systems are constantly studying different alternatives to increase efficiency and reduce cost. In this context, vehicle routing problems are of particular interest and several applications arise from the waste collection sector.

Regarding packaging waste, there are mainly two types of selective collection available in Portugal: drop-off containers or curbside (door-to-door). In the drop-off system, the citizens/consumers have to move to a container nearby their home to drop the separated materials; in the door-to-door system the recyclable containers are inside their home building and are collected at a predetermined day of the week. The main system used in Portugal is the drop-off containers. The three types of recyclable packaging materials (paper, glass and plastic/metal) are dropped by the consumers into special containers. Then, those materials are collected by the company responsible for the waste collection system of that area on a regular basis. The drop-off containers are mainly collected by a top loaded truck, with a single compartment. Therefore, single-material routes are performed, what implies that the same collection site is visited three different times to collect each material individually. One alternative to this kind of system is to use vehicles

© Springer International Publishing Switzerland 2015   131
A.P.F.D. Barbosa Póvoa and J.L. de Miranda (eds.), *Operations Research and Big Data,*
Studies in Big Data 15, DOI: 10.1007/978-3-319-24154-8_16

with compartments and collect two or even the three materials simultaneously without mixing them.

This study was motivated by a real recyclable waste collection system operating in Portugal that aims to assess the impact on the distance travelled by using vehicles with multiple compartments to collect the drop-off recyclable containers instead of using single compartments vehicles. For that, a heuristic approach is developed to solve the multi-compartment vehicle routing problem and applied to the real case study. The results regarding the distance travelled are compared with the current solution, where single-material routes are performed.

The paper is structured as follows: in section 2, a literature review on the Multi-Compartment Vehicle Routing Problem (MCVRP) is presented. The case study is described in section 3. In section 4 we propose a heuristic approach to solve the MCVRP. The results are presented and discussed in section 5. Lastly, in section 6 the conclusions are presented.

2   Literature Review

In the Capacitated Vehicle Routing Problem (CVRP) the customers request the delivery (or collection) of just one product. A homogenous vehicle fleet is based at a central depot and the optimal delivery or collection routes must be designed, respecting the capacity of the vehicles. In the Multi-Compartment Vehicle Routing Problem (MCVRP) the customers request the delivery (or collection) of different products which cannot be commingled during transport. For that, vehicles with multiple compartments are used to co-transport the products. In this case, the compartments capacity cannot be exceeded.

Despite several real life applications (namely in distribution of food and petrol), the MCVRP has not been studied extensively. However, some works have been recently published. El Fallahi et al. (2008) formulated the MCVRP and proposed three algorithms to solve it: a constructive heuristic, a memetic algorithm and a tabu search method. The authors adapted well-known instances for the classical VRP to test the methods proposed. Muyldermans and Pang (2010) presented a local search procedure for the MCVRP and have tested on CVRP instances and MCVRP instances. The authors also generated new instances to assess the benefits from using vehicles with multiple compartments on waste collection field. They compared the routing cost for co-collection (solving a MCVRP) with the routing cost for separate collection (solving a CVRP) and concluded that co-collection is better when the number of product to collected is higher, when the vehicle capacity increases, when the products are less bulky, when more clients request all commodities, when client density is lower and when the depot is more centrally located in the collection area. Derigs et al. (2011) developed and implemented a portfolio of different heuristic components for solving the MCVRP and tested on literature instances. Reed et al. (2014) proposed an ant colony algorithm and tested for the MCVRP with two compartments. Coelho and Laporte (2015) proposed a classification scheme for the fuel distribution problem regarding the ability to split the content of a compartment between several deliveries and the ability to split a customer tank to receive deliveries from different vehicles. The authors stated that only the unsplit-unsplit problem is traditionally treated in the literature and the three other cases (split compartments-split tanks, split compartments-unsplit tanks, unsplit compartments-split tanks) are new and are modelled and solved for the first time in that work. A real-life application is presented by Lahyani et al. (2015). The authors tackled the olive oil collection problem in Tunisia, where three different grades must be kept separate during transportation. An exact branch-and-cut algorithm is proposed to solve the problem.

In face of this literature review, we can conclude that the MCVRP is getting more attention from the academia in the recent years but only literature instances have been tackled (with the exception of the work of Lahyani et al., 2015) . Real-life applications are seldom studied. Therefore, this work explores this opportunity and studies an application of the MCVRP in the collection of recyclable materials.

3   Case-Study

In Portugal there are 30 recyclable waste collection systems. One of them is Valorsul, which is responsible for the selective collection in 14 municipalities located at West Region. Valorsul owns a vehicle fleet of 12 vehicles that is based at one depot located in the municipality of Cadaval. All vehicles have one compartment with 20 m^3, meaning that single-material routes are performed. Material glass is collected by a top loaded truck with no pressing function. Paper and plastic/metal are collected also by top loaded trucks, but with pressing function. The vehicle crew includes one driver and one assistant. There are 82 routes established: 26 for paper collection, 26 for plastic/metal collection and 30 for glass collection. Considering the collection routes performed between January and September 2013, Table 1 shows the average time between collections, number of containers collected by route, distance travelled per route and amount of material collected per route for each type of recyclable material.

Table 1 Current indicators for Valorsul routes performed between January and September 2013 (average values)

Recyclable material	No. routes performed	Time between collections (days)	No. containers collected per route	Distance travelled per route (km)	Amount collected per route (ton)
Paper	816	9,3	82	135,3	2.8
Plastic/Metal	740	8,3	81	137,9	2.1
Glass	342	20,5	83	151,2	9.8

Paper and plastic/metal have similar indicators meaning that they could be collected together on the same route, if vehicles with two compartments were used. Glass differs greatly from the other materials regarding time interval between collections and the amount collected per route (given its high density compared with the other two materials). Therefore, the new approach that the company wants to test will consider the use of vehicles with two compartements to collect simultaneously paper and plastic/metal. These vehicles will have the same volume capacity than the ones actually used by the company (i.e. 20 m^3), that is going to be split accordingly to the density and weight produced of the two materials (see Section 5.1). It is considered that both compartments have pressing function. The aim of this work is to assess the benefits (in terms of distance to be travelled) of using multi-material routes instead of the traditional single-material routes (see Figure 1).

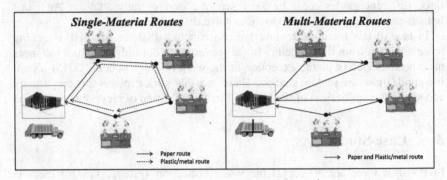

Fig. 1 Illustration of single-material routes *versus* multi-material routes

The two materials chosen to be collected together have different material density. At the containers, paper density is about 40 kg/m^3 and plastic/metal 20 kg/m^3; with vehicles with pressing function, densities increase to 250 kg/m^3 and 150 kg/m^3, respectively. These values were given by Valorsul. On the other hand, the annual amount collected of these two materials is also different. In 2012, Valorsul collected 15309 ton of paper and only 8583 ton of plastic/metal, meaning that the container's filling rate of each material is different. Given that the materials have different densities and different container's filling rates, there are two issues that must also be addressed when planning the collection routes with vehicles with multi-compartments: *(i)* how to define each compartment capacity and *(ii)* when to collect both containers.

4   Heuristic Approach to Solve the MCVRP

We propose a heuristic to solve the MCVRP based on the approach "cluster first-route second". When defining clusters, the container's filling rate, compartments capacity for each material, and the distance between the containers are taken into account. One of the main features of recyclable waste collection problems is that the containers have different filling rates among themselves due to its location. To tackle that, the main idea is to define large clusters with containers with low filling rates and small clusters with containers with high filling rates. Low filling rates mean that those containers have to be collected with a longer time interval between collections. Therefore, that cluster has to be visited fewer times during a year. On the other hand, clusters with containers with high filling rates have to be visited more frequently, thus, more times during a year. The idea is to have clusters with containers with similar collection frequencies since they are going to be collected in the same route.

The cluster phase starts by choosing a random site as a seed point and then more sites will be included to that cluster within a given radius of R km. For the sites included in that circle, the amounts to collect of the two materials in each site are summed and ranked in a descending order. Given this ranked list, the heuristic starts adding containers to the first cluster until no sites are available within the circle or one of the capacity of the vehicle compartments is exceeded. Then, another site is chosen as seed and the process is repeated until all sites are included in clusters. The radius value is a parameter and it is not known the best value to use. Therefore, several radius values will be tested. Moreover, as the seed points are randomly chosen, different solutions will be generated.

As a result of the clustering phase, we end up with n TSPs that need to be solved. We start by applying the savings algorithm (Clarke and Wright, 1964) followed by two local search procedures: two-point-move and 2-opt.

Given the interdependency between the filling rate of a container and the time interval between consecutive collections, it was considered an initial value for the time interval between collections in order to define the clusters and routes. After routes are defined, the time interval between collections (T) is maximized in order to decrease the number of times each route has to be performed in a year. To maximize that value it is taken into account the daily disposal rate of each container for each material (d_{im}), the vehicle compartments capacity for each material (V_m) and the containers capacity (C_m). For each route with n containers, the following problem is solved:

$$Max\ T \tag{1}$$

$$\sum_{i=1}^{n} d_{im}T \leq V_m, \quad \forall m \tag{2}$$

$$d_{im}T \leq C_m, \quad \forall i, \forall m \tag{3}$$

$$T\ integer \tag{4}$$

The first phase of the heuristic (clusters definition) was coded in VBA. The second phase (route definition) was solved using the VRPH Library implemented in C++ available in the work of Groer (2008). The third phase (maximization problem) was solved using MS Excel Solver.

5 Application to the Case-Study

5.1 Data Collection

Given the vast intervention area of Valorsul (more than 3000 km^2) and the high number of containers (7807 containers), it was decided to test the use of multi-compartments vehicles in a small area first (test-area). The test-area was defined taking into account the routes performed by the company. We select three routes for paper and three routes for plastic/metal considering that: (i) the routes should be close to each other to obtain a contiguous test-area, and (ii) the routes for each material should be as similar as possible, given the number of containers to collect and the time between collections, in order to be possible to implement a joint collection. The selected routes are characterized at Table 2.

Table 2 Characteristics of the selected routes for the test-area

Selected Routes	Average distance (km)	No. of containers	Average time interval (days)	Average weight collected (kg)	Average daily amount collected by container (kg/container.day)
#4 Paper	106	107	6,02	4958	7,7
#4 Plastic/metal	106	108	5,98	2885	4,5
#7 Paper	87	92	7	5127	7,9
#7 Plastic/metal	86	87	5,56	2546	5,3
#13 Paper	130	90	7,6	4494	6,6
#13 Plastic/metal	128	90	7,03	2564	4,1

After selecting the test-area, we need to determine the amount to be collected in each collection site and the time between collections. For that, it was previously computed the average daily amount collected by container. This estimation was based on the time interval between two consecutive collections and on the average amount collected by container in each route. These results are shown also at Table 2. We consider the smallest time interval between two consecutive collections as the time interval to compute the amount to collect in each collection site. Therefore, we use the value of 5 days, but after the multi-material routes were defined, this value will be maximized for each route as explained at Section 4.

To decide the compartments capacity, we compute the ratio between the amount collected of paper and the amount collected of plastic/metal. We conclude that for each kg of paper collected, 0.62 kg of plastic/metal are collected. Given the density of both materials in the vehicles (250 kg/m^3 for paper and 150 kg/m^3 for plastic/metal) and the vehicle capacity (20 m^3), the vehicle should have 9.8 m^3 allocated to collect paper and 10,2m^3 to plastic/metal. This means that the compartments capacities are 2450 kg for paper and 1530 kg for plastic/metal.

5.2 Results

Given that the heuristic developed has a random element (the seed choice to create the clusters), we generate five solutions and present the average values obtained. Regarding the parameter radius, four different values were tested (20 km, 25 km, 30 km and 35 km).

Seven clusters/routes were defined to collect paper and plastic/metal containers within the test-area. Table 3 shows that the first clusters have more containers since the heuristic aggregate first the containers with lowest filling rates, thus more containers are visited in the first routes and less in the last routes.

At Figure 2 it is shown the usage rate of each compartment for each route of solution 5 with radius 20 km. It can be seen that until route 4 we have high usage rates for both compartments (higher than 64%) and for the last three routes we have some cases lower than 50% and a very lower rate for route 7.

Regarding distance, the best solutions found for each radius are presented at Table 4. However, at this point it is not possible to conclude what is the best solution considering annual distance because the time interval between collections has to be analyzed.

Table 3 Average number of containers collected per cluster/route

Clusters/Routes	Radius 20 km	Radius 25 km	Radius 30 km	Radius 35 km
1	71,0	71,6	77,2	75,6
2	55,6	58,0	58,8	51,8
3	52,6	57,4	48,4	47,2
4	41,4	37,6	39,4	56,6
5	42,0	31,6	34.6	29,2
6	23,6	30,2	18,8	25,0
7	2,8	2,6	11,8	3,6

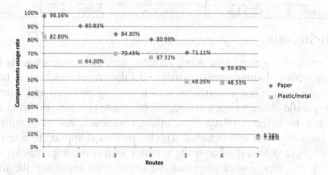

Fig. 2 Compartments usage rate for each route (radius=20km, solution 5)

Table 4 Best solution for each radius value

	Radius 20 km	Radius 25 km	Radius 30 km	Radius 35 km
Best solution	Solution 5	Solution 2	Solution 5	Solution 1
Distance (km)	510,7	509,3	514,6	530,3

It was considered as an input data a time between collections of 5 days. However, this value could now be maximized given the compartments usage rates and the containers filling rates. If the compartments usage rate is less than 100% and all containers collected have a filling rate lower than 100%, that route could be performed with a 6 days interval, for example. Table 5 shows the time between collections maximized. For example, in the best solution with radius 20 km, two routes have increase the time between collection to 6 days and in one route that value increase to 7 days. This will have a great impact on the annual distance travelled. In fact, the best solution considering only the distance of each route (Table 4) is not the best solution considering the annual distance. At Table 5 it can be seen that the best solution is the one with radius of 20 km.

The current solution for the test-area implies six balanced routes in terms of number of containers collected, with a time between collection of 5.5 days to 7.6 days (see Table 2) and a total distance travelled per year of 35968 km. This means that performing multi-material routes decreases the total distance travelled in 5% (34273 km *vs.* 35968 km), but more routes are performed (7 routes *vs.* 6 routes) and they are more unbalanced regarding the number of containers to collected.

Table 5 Time between collections for each route belonging to the best solution for each radius

	Radius 20 km	Radius 25 km	Radius 30 km	Radius 35 km
Best solution	Solution 5	Solution 2	Solution 5	Solution 1
Route 1	6	5	5	5
Route 2	5	5	5	7
Route 3	5	6	5	5
Route 4	5	7	5	5
Route 5	6	5	7	5
Route 6	7	5	5	5
Route 7	5	5	5	5
Annual Distance	**34273 km**	34946 km	36286 km	37248 km

6   Conclusions

This paper assesses the impact of using vehicles with two compartments to collect paper and plastic/metal simultaneously using real data from Valorsul. A heuristic is proposed to solve a multi-compartment vehicle routing problem and attention was given to the specific characteristics of a recyclable waste problem: different collection frequencies due to different filling rates among containers, materials with different densities at the containers and vehicles. The heuristic was applied to a test-area and savings of 5% were obtained when using vehicles with two compartments instead of one. However, the new routes proposed are unbalanced regarding the number of containers to collect, and this represent a drawback to the implementation of this solution. To overcome this drawback and as further work, the cluster phase should be improved in order to get all clusters with a greater and balanced number of containers to collect. Narrowing the radius value will contribute to solve this problem. This study should also be complemented with a feasibility study to assess the feasibility of the investment in new vehicles or in adapting the vehicles to have two compartments.

References

Clarke, G., Wright, J.W.: Scheduling of vehicles from a central depot to a number of delivery points. Oper. Res. 12, 568–581 (1964)

Coelho, L.C., Laporte, G.: Classification, models and exact algorithms for multi-compartment delivery problems. Eur. J. Oper. Res. 242, 854–864 (2015)

Derigs, U., Gottlieb, J., Kalkoff, J., et al.: Vehicle routing with compartments: applications, modelling and heuristics. OR Spec. 33, 885–914 (2011)

El Fallahi, A., Prins, C., Wolfler Calvo, R.: A memetic algorithm and a tabu search for the multi-compartment vehicle routing problem. Comput. Oper. Res. 35, 1725–1741 (2008)

Groer, C.: Parallel and serial algorithms for vehicle routing problems, PhD Thesis, University of Maryland (2008)

Lahyani, R., Coelho, L.C., Khemakhem, M., et al.: A multi-compartment vehicle routing problem arising in the collection of olive oil in Tunisia. Omega 51, 1–10 (2015)

Muyldermans, L., Pang, G.: On the benefits of co-collection: Experiments with a multi-compartment vehicle routing algorithm. Eur. J. Oper. Res. 206, 93–103 (2010)

Reed, M., Yiannakou, A., Evering, R.: An ant colony algorithm for the multi-compartment vehicle routing problem. App. Soft. Comp. 15, 169–176 (2014)

Ramos, T.R.P., Gomes, M.I., Barbosa-Póvoa, A.P.: Assessing and improving management practices when planning packaging-waste collection systems. Res. Cons. Rec. 85, 116–129 (2014)

Pricing for Internet Sales Channels in Car Rentals

Beatriz Brito Oliveira[1], Maria Antónia Carravilla[1], José Fernando Oliveira[1], Paula Raicar[2], Delfina Acácio[2], José Ferreira[2], and Paulo Araújo[2]

[1] INESC TEC and Faculty of Engineering, University of Porto, Porto, Portugal
[2] Guerin Car Rental Solutions, Lisbon, Portugal

Abstract. Internet sales channels, especially e-brokers that compare prices in the market, have a major impact on car rentals. As costs are heavily correlated with unoccupied fleet, occupation considerations should be integrated with swift responses to the market prices. This work was developed alongside Guerin, a Portuguese car rental, to build a tool that quickly updates prices on e-brokers websites to increase total value. This paper describes the specificities of the problem and their implication on the solution, and presents an adaptative heuristic to update prices and the system's architecture.

1 Introduction

Similarly to other tourism-related sectors, the car rental business has been deeply impacted in the past years by internet sales, namely by the development of a new sales channel: broker websites that compare the offers of different competitors. The impact of this channel is especially relevant for car rentals due to the lack of differentiation of their product, if compared, for example, with the hotel business. As the vehicles and pick-up stations are the same and the clients are able to compare all the offers in the market with full transparency, the price takes an even more determinant role in their decision.

The ultimate goal of every company is to maximize its revenue. For car rental companies, the main slice of the costs is related with unoccupied fleet. Therefore, the "revenue challenge" deals not only with uncertain demand (highly dependent on the companies' positioning versus the other prices on the market); it also deals with the need to maximize the occupation of the fleet for each day, ensuring the cars were booked at the highest possible price. Moreover, for the e-brokers channel, it is of the utmost importance to be watchful and agile in order to respond to changes on the market prices. This requires processing a massive amount of data in an extremely short time-frame.

This paper presents the work developed in *Guerin Car Rental Solutions*, a Portuguese car rental company, to build a tool that allows for a swift, systematic,

A.P.F.D. Barbosa Póvoa and J.L. de Miranda (eds.), *Operations Research and Big Data*, Studies in Big Data 15, DOI: 10.1007/978-3-319-24154-8_17 regular and profitable update of all the company's pricing positions in the market. This tool is based in an heuristic procedure that is adaptative in the way it continuously corrects the prices responding to the changes in the market conditions (demand and competitors' prices), in order to attract the right (number of) customers at each point in time and thus increasing the value collected for the fleet each day.

2   The Problem

2.1   Brief Description

The main decision of this problem is setting the price to charge for a specific *search* that a customer makes online a certain number of days in advance (e.g. 30 days beforehand). In fact, this problem is highly dependent on the *antecedence* of the search versus the start of the reservation. The e-brokers channel is typically used by the *leisure* segment, allowing prices to vary over time (this is not true for the corporate segment, for example, where other reservation formats, such as pre-established contracts, may limit the price variation). Therefore, there is a need to balance the goal of occupying the fleet and the goal to do so at the highest possible price.

A search is characterized by the e-broker website where it is made, the starting date (vehicle pick-up) of the reservation, the rental length (in days), the pick-up station, and the type of vehicle required. The *number of days in advance* is calculated based on the starting date of the reservation (see Figure 1).

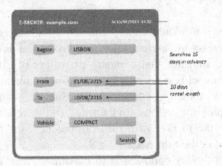

Fig. 1 Example of a seach

The goal is to develop a tool that is able to calculate, in short intervals (e.g. every two hours) the prices to set for each *search* the customers can make.

2.2   *Important Characteristics*

One of the characteristics of this problem that influences the most the pricing decision-process is the *transparency* between competitors. When searching online on the e-brokers website, the customers search for a specific vehicle, location and dates and the results are retrieved for all competitors with an offer in the market, usually organized by price. This leads to a *best deal* effect, in which the competitor offering the lowest price gets the most attention, even if the price difference to the others is marginal. This can only be surpassed by some competitors, whose considerable market share and customer awareness can trigger customer preference even if their price is not the lowest in the market.

However, one can note that although the customers have full transparency between competitors, their searches are usually focused on specific dates and locations, *hindering the transparency between prices of the same company*. For example, if a customer wishes to rent a car for a leisure trip, he/she is not likely to change the trip dates (or the region of the car pick-up) in order to get a cheaper deal (this is not necessarily true for other businesses such as airline). Therefore, there is a *flexibility* for the car rental to set the prices of different searches independently of each other.

Nevertheless, it is important to bear in mind that the prices of the different searches are not completely independent: 1) the main slice of the costs of car rental companies derives from unoccupied fleet; 2) fleet occupation and price influence each other; low prices increase the pace at which new reservations are made, increasing the pace of fleet occupation (and high prices have the opposite effect) - it is thus possible to use the prices to accelerate/decelerate occupation rate; 3) a fleet can be occupied by a myriad of different searches (and thus prices) - in fact, the fleet of a certain vehicle type in a certain region is, in a certain day, occupied by reservations made in different e-brokers that started in different days and stations and will have different durations. The main challenge faced was thus related to the amount of data to process in order to calculate the price for every *search*, in a short period of time (two hours), including processing the current prices in the market for each competitor for each search, and the occupation of the fleet(s) affected. For the company considered in this paper, the amount of different searches for which updated prices must be calculated at each iteration is in the range of the tens of thousands.

2.3 Literature Review

This problem is herein regarded as a *revenue management* problem. In [1], revenue management is related with three different types of ``demand-management decisions": structural decisions (related with the configuration of the sale -- e.g. auction), price decisions (related with the price to set for different products, customers, product life-cycle, ...) and quantity decisions (e.g., related with the allocation of resources to segments). The authors state the importance of recognizing the business context so as to understand the relevance of price and quantity flexibility, which will have a deep impact on revenue management strategies. For example, in the hotel business, it may difficult to increase capacity of rooms whilst prices are significantly easier to change. In fact, in the problem tackled in this paper, the main focus is on *price decisions*, as the flexibility to change prices is significantly superior to the one to change capacity. In [2], business characteristics that justify why companies adopt revenue management programs are reviewed. It is possible to verify that the problem herein described presents these characteristics. Firstly, as in most service-oriented problems, excess resources are impossible to store (in this case, if a car is not used some day, the capacity is lost). Also, pricing decisions are made with uncertain demand, and different customers with different willingness to pay have different demand curves while sharing the same resources. Moreover, the company is profit-oriented and able to freely implement the decisions.

The fact that this problem is related with a web-based sales channel has a deep impact on the problem definition. Already in 1998, it is argued that internet-based marketplaces decrease the customer cost to obtain information and compare offers, which promotes price competition; moreover, they increase the ability of the seller to charge different prices to different customers (or to charge different prices over time), which reduces customer surplus and increases company's profit ([3]).

Revenue management is historically linked with the airline business. In [4], the implementation of a revenue management system at Western Airlines is described, with a seat inventory control based on the Expected Marginal Seat Revenue (EMSR) decision model, which sets and revises booking limits to the number of seats available at different prices. This model takes into account the uncertainty of demand and is based on the value of the expected revenue per seat per class or segment, thus defining their protection levels (number of seats hold for sale for certain segments) and, consequently, booking limits. Over the time, revenue management has been applied to different sectors and situations. In [5], the maximization of revenue in a network environment is tackled by defining dynamic policies for the allocation of shared resources to different types of uncertain demand. The authors propose a solution based on approximate dynamic programming, which may have applications on airlines, hotels, car rentals, amongst other businesses.

Revenue management on car rental business was early on tackled by [6], who described the implementation process of the Yield Management System (YMS) in Hertz, designed to help decisions related with pricing, fleet planning and fleet deployment between stations. Hertz's YMS segmented the market with different-valued offers and helped decide when to make these offers. It also protected fleet for higher-value reservations when supply was short. Also in [7] the revenue management program of a different company is described - National Car Rental. This system was designed to *manage capacity*, tackling fleet planning issues, planned upgrades (with inventory protected using the above-mentioned EMSR model), overbooking, and creating a *Reservations Inventory Control* procedure for selection of the most profitable reservations amongst similar ones with different lengths of rent. It also tackled *pricing issues*: for different segments, the model recommended prices to maximize occupation based on a target utilization. This model for pricing, which inspired the heuristic proposed, is composed of two parts. Firstly, an elasticity model is created relating historic rates with variations in demand. Then, the comparison between occupation (or demand) *forecast* and *target* leads to the required change in booking pace. From the elasticity model it is thus possible to retrieve the rate that will induce the required change in the pace of demand.

More recently, a revenue management problem in a car rental network is tackled in [8]. The authors propose a stochastic programming approach to optimize fleet deployment between locations and capacity controls to protect fleet for higher revenue reservations, under a demand of probabilistic nature; in this context, the prices are not considered as decisions. Other works apply revenue management techniques to the car rental context, such as [9]. Here, the authors aim to allocate vehicles of different categories to different customer segments, following a framework on which the company must decide if it is more profitable to accept a rental request or not and thus define acceptance policies. A parallel can also be found with pricing problems. For example, in [10] dynamic pricing strategies are studied, which might be applied to the car rental context. In this work, a common formulation for allocation of capacity and dynamic pricing is presented.

3 The Proposed Solution

The proposed solution to set the prices for the e-brokers sales channel is a heuristic designed to be swift, adaptative to the market and fleet conditions, and fast to implement. Research on revenue management and pricing in car rental is growing significantly on the stream of optimal policies and controls for allocating the right customers to the right-priced offers. However, to the best of the authors' knowledge, there is still a gap on the development of methods that address the link between the need to protect fleet and the need to take in consideration competitors' prices and the company's competitive position. This was a specific requirement of the work herein developed.

This heuristic is based on the concept of *goal occupation*, which is described in this chapter. A full decision tool / system was also designed and implemented and is also described in this chapter, representing the functioning conditions surrounding the heuristic procedure. This tool allows the car rental company to parametrize the heuristic procedure based on its business sensitivity, as well as to monitor key fleets and seasons of the year, providing useful indicators related to regional fleet balancing, *real* margins applied by e-brokers and market price fluctuations.

3.1 Heuristic Procedure Based on Goal Occupation

The main objective of this problem is to set the prices to charge for each search, at each time distance to the reservation, so that the revenue of the car rental company is maximized. For that, one needs only to ensure that the capacity of each fleet is only booked to the maximum of its capacity by the different types of searches that influence it. However, the relation between the price to charge, the minimum price in the market for the same search, and the amount of reservations that one is able to get from it is intrinsically hard to define realistically. Therefore, the concept of *goal occupation curve* is introduced: the percentage of vehicles from a fleet that the company aims to have occupied with reservations for a certain date, at a certain time distance. For example, the company wants to have enough reservations to occupy 60% of fleet F on day D, 30 days before day D (see Figure 2).

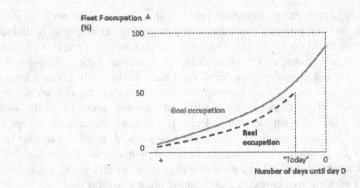

Fig. 2 Representation of the evolution of goal and real occupation of a fleet for a specific day

Following this goal occupation curve allows the company to increase the value attained for the fleet sold. That is to say that the company is able to define how much vehicles to hold for the "late" clients that are willing to pay more for them. This concept is parallel to the *target utilization* [7]. The heuristic proposed is thus based on the concept introduced in [7] that states that changes in the rates influence the difference between target and forecast occupation, which should be minimized. In our problem, as this procedure is designed to be run progressively and frequently (every two hours) and to be adaptive, actual occupation is used (not forecast). Therefore, in this heuristic, the prices charged are a mean to minimize the distance between the observable fleet occupation and the *goal occupation*, for a certain time distance to the reservation. Based on the discussion regarding the problem characteristics, it is assumed that companies are only able to get clients if their price is the lowest in the market. Therefore, if the real occupation is lower than the goal, the price should be set to be the lowest in the market (although never allowing for the price to fall below a certain minimum price). Conversely, when real occupation is higher than the goal, the company should increase the price in order to reduce occupation rate, yet striving to achieve the desired pricing position versus competitors. If the occupation becomes extremely high, however, the increase should become independent of the competitors' prices, so as to hinder more (undesired) occupation.

Real occupation is calculated based on the concept of *most constrained day*. In fact, a certain *search* whose price is being settled will be translated in a reservation that may last for more than one day. In order to be conservative, all calculations consider the most occupied day in the reservation. Also, to calculate the new price is critical to "see" the prices the customer "sees" for each player. Therefore, when deciding the price to settle, the *margin* the e-broker is applying on the company's price should also be added.

This procedure settles the price for each search independently of the other searches, although several searches influence the same fleet occupation, as seen in the discussion above. The only searches that are linked are the ones that share all characteristics except for the group, and whose hierarchy of the groups was previously set (e.g. the price of a compact car should always be lower than the price of a luxury model). The searches are thus mostly considered independent in order to agilize the procedure since its main consequence is a higher degree of conservatism. For example, if a certain fleet in under-occupied, this heuristic will decrease the price for all searches that influence this fleet (as if they were the only ones affecting it). The effect may be higher than expected due to this "over-kill". However, as this procedure is adaptative, if this happens, in the next run the fleet will be slightly over-occupied and the prices will be adapted to this new context. In fact, this *easiness to adapt* is one of the most important characteristics of this procedure.

3.2   Full System Overview

Figure 3 aims to describe the working flows of the full system designed. The system is divided in two areas: a User Area and a Work Area. The first is where the heuristic is parametrized, as well as the frequency to run each search. It also allows the user to choose which key groups to monitor. The second is where the connection with the main input and output systems is established and the procedure itself is based.

There are four main types of inputs to this system. The first two types of inputs are provided by external systems while the latter are user-defined:

- The *prices* currently available for the customers on each e-broker website, for the company and all its competitors;
- The *occupation* of the fleets for each day in this horizon;
- The main parameters of the system;
- The key groups to monitor.

Every pre-defined interval of time (hourly or every two hours), the system follows three steps. Previously, the user has defined the schedule of each search, based on its characteristics. For this, the user may define several command lines, such as "recalculate prices for all searches/reservations starting next month" or "for all that take place on 2015 High Season". From this, the first step of the system is to *list all the searches for whom to recalculate prices at this moment*. Secondly, the heuristic procedure recalculates prices for all the listed searches and sends them to the e-brokers. Finally, for the key monitoring groups, the main indicators are made available to the user.

It is important to note that companies do not have much control over the margins applied by e-brokers and thus a potential margin is computed and used in step 2: this margin is equal to the last company's price retrieved from the e-brokers' website divided by the recalculated price sent to the e-broker on the last iteration.

For some pre-selected key groups, the evolution of the occupation and price/margins is also monitored. For example, the user may follow the evolution of a certain fleet's occupation for a specific week and compare different regions through interactive graphic displays.

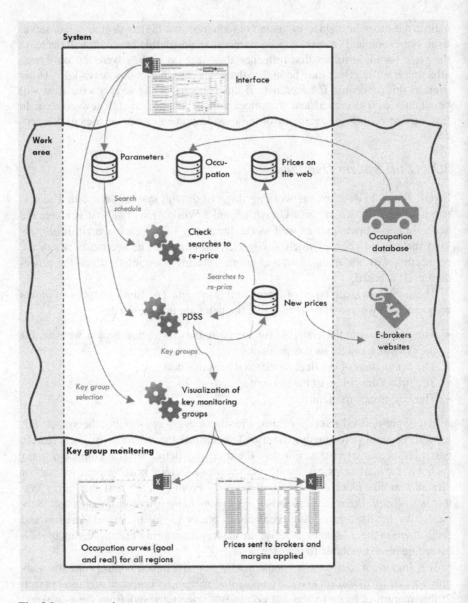

Fig. 3 System overview

4 Conclusions and Future Work

This system, currently being implemented on *Guerin*, will enable the company to adapt its prices to the occupation of the fleet and the fluctuations of the market, seizing significantly more value from the customer willingness to pay. Moreover, it will be possible to monitor key periods of the year or key vehicle groups, as well as the margins added to the prices by the e-brokers.

The future work lays ahead in two different dimensions. On the business side of the problem, there is a need to measure the actual impact of this system and, based on that, refine key parameters of the problem such as the *goal objective curve*. This is possible since this system is designed to save all the data for future research.

On the academic side of the problem, since this is, at the best of the authors knowledge, yet to be fully explored by the academic community, future work is needed to build models that are able to bring even more value to the company by maximizing the revenue and tackling all the prices that relate to the same fleet occupation in an integrated way. Moreover, as these models will lead to (realistic) large instances, solution methods to solve them will be required. In a second stage, it will also be interesting to include *fleet sizing and deployment* issues in the problem, integrating decisions of vehicle transfer between regions when occupations are unbalanced.

References

[1] Talluri, K.T., Van Ryzin, G.J.: The theory and practice of revenue management. Springer Science+Business Media (2006)

[2] Netessine, S., Shumsky, R.: Introduction to the Theory and Practice of Yield Management. INFORMS Transactions on Education 3(1), 34–44 (2002)

[3] Bakos, Y.: The emerging role of electronic marketplaces on the Internet. Communications of the ACM 41(8), 35–42 (1998)

[4] Belobaba, P.P.: OR Practice - Application of a Probabilistic Decision Model to Airline Seat Inventory Control. Operations Research 37(2), 183–197 (1989)

[5] Bertsimas, D., Popescu, I.: Revenue Management in a Dynamic Network Environment. INSEAD Working Paper Series 2000/64/TM (2000)

[6] Carroll, W.J., Grimes, R.C.: Evolutionary Change in Product Management: Experiences in the Car Rental Industry. Interfaces 25(5), 84–104 (1995)

[7] Geraghty, M.K., Johnson, E.: Revenue Management Saves National Car Rental. Interfaces 27(1), 107–128 (1997)

[8] Haensel, A., Mederer, M.: Revenue management in the car rental industry: A stochastic programming approach. Journal of Revenue and Pricing Management 10(1), 290 (2011)

[9] Guerriero, F., Olivito, F.: Revenue Models and Policies for the Car Rental Industry. Journal of Mathematical Modelling and Algorithms in Operations Research 13, 247–282 (2014)

[10] Maglaras, C., Meissner, J.: Dynamic pricing strategies for multiproduct revenue management problems. Manufacturing & Service Operations Management 8(2), 136–148 (2006)

Eco-efficiency Assessment at Firm Level: An Application to the Mining Sector

Renata Oliveira, Ana Camanho, and Andreia Zanella

Faculdade de Engenharia da Universidade do Porto, Portugal

Abstract. Assessing firms' Eco-efficiency is important to ensure they succeed in creating wealth without compromising the needs of future generations. This work aims to extend the Eco-efficiency concept by including in the assessment new features related to environmental benefits. Eco-efficiency is evaluated using a DEA model specified with a Directional Distance Function. The new methodology proposed in this paper is illustrated with an application to world-class mining companies, whose results and managerial implications are discussed.

Keywords: Eco-efficiency, Directional Distance Function, Mining Companies.

1 Introduction

The Eco-efficiency concept has gained momentum during the 1990s, with the release of two seminal publications: "Environmental rationality" (Schaltegger and Sturm, 1990) and "Changing Course" (Schmidheiny, 1992). As a result, eco-efficiency issues began to figure prominently in the scientific fields of Sustainable Development and Business Management (e.g., Choucri, 1995; Cramer, 1997; Esty and Porter, 1998). In the transition to the 2000s, the World Business council of Sustainable Development (WBCSD) has described Eco-efficiency at the firm level as follows:

> *"keeping the business competitive while reducing material and energy requirements, minimizing the dispersion of toxic wastes and maximizing the sustainable use of renewable resources"* (WBCSD, 2000, p.10).

The literature includes a variety of studies that intend to quantitatively reflect the Schmidheiny's definition in contexts involving several indicators, such as Glauser and Müller, 1997; Kortelainen and Kuosmanen, 2007; Zhang et al., 2008; Picazo-Tadeo and Prior, 2009. These applications have expressed Eco-efficiency by ratios of the economic achievements to the environmental burdens associated with wealth generation. In the presence of multiple indicators, these have to be aggregated using predefined or optimized weights. The studies available in the literature do not cover entirely the WBCSD criteria for assessing Eco-efficiency at the firm level, specially concerning the sustainable use of renewable resources.

© Springer International Publishing Switzerland 2015

A.P.F.D. Barbosa Póvoa and J.L. de Miranda (eds.), *Operations Research and Big Data,*
Studies in Big Data 15, DOI: 10.1007/978-3-319-24154-8_18

The first contribution of this paper is to provide a comprehensive view of firms Eco-efficiency by using an ehanced range of indicators. These indicators are aligned with the Eco-efficiency criteria of WBCSD and are classified in three dimenions: economic benefits (e.g., value-added), environmental benefits (e.g., use of renewable resources) and environemntal burdens (e.g., emissions of pollutants). The second contribution of this paper is the development of an enhanced model, based on a directional distance function (DDF), which proposes simultaneous adjustsments to indicators from different categories. The third contribution consists of the application of the model to world-class mining firms. Mineral exploitation can be considered one of the most environmentally impacting economic activities in the planet, so the assessment of good environmental practices in this sector is clearly an important issue.

This paper is organized as follows: section 2 describes the enconomic and environmental indicators used for assessing Eco-efficiency of world-class mining companies. Section 3 presents the methodology of the enhanced Eco-efficiency assessment and describes the new model specified with a DDF. Section 4 discusses the results and section 5 concludes the paper.

2   Indicators Used for Eco-efficiency Assessment

The indicators selected for the Eco-efficiency assessment of mining companies, reported on Table 1, are aligned with the 10 principles of the International Council on Mining and Metals (ICMM). The ICMM sectorial initiative aligns with important international sustainability guidelines, such as ISO (2003) and GRI (2013).

Table 1 Indicators used for Eco-efficiency assessment

Dimension	Category	Indicator
Economic	Benefits (y_r)	Economic value-added (y_1)
Environmental	Benefits (h_q)	Renewable energy use (h_1)
		Recycled materials use (h_2)
		Water recycled or reused (h_3)
		Conservation areas supported (h_4)
		Investments on environment (h_5)
	Burdens (p_k)	Non-renewable energy use (p_1)
		Withdraw water use (p_2)
		Waste produced (solid and liquid) (p_3)
		Air emissions (GHG and pollutants) (p_4)
		Volume of significant spills (p_5)
		Total environmental fines (p_6)

For the economic dimension, economic benefits are related to firms' wealth generation. These are represented by the annual economic value-added.

The environmental dimension of this framework includes two categories of indicators: the burdens to be minimized (e.g. air emissions) and the benefits to be maximized (e.g. conservation areas supported). In the mining sector, waste generation is an issue of major concern. The amounts of mining waste generation depend on the quality of the extraction and transportation processes as well as on the ore contents in soil. This study only considers wastes from packaging, raw and hazardous materials, to ensure data comparability across firms.

The list of companies analyzed and the corresponding values of the indicators used for the Eco-efficiency assessment are available in the appendix (Table A.1 and A.2). The companies studied hold several mines around the world and may exploit various types of ores (e.g., Vale (U1) is a multinational company that exploits primarily metallic ores and operates in ten different countries). Despite the firms' different mix of products, they should observe the same international standards for environmental and economic performance, so their comparison based on the indicators reported on Table 1 is legitimate.

3   The DEA-Based Model for Extended Eco-efficiency

The quantification of the extended Eco-efficiency measure was accomplished using a Directional Distance Function (DDF) model, first proposed by Chambers et al. (1996). The formulation shown in (1) follows the approach adopted by Färe et al. (2014), which involves an equal treatment of all indicators representing burdens to be minimized, irrespectively of their intrinsic nature being an input (e.g., withdraw water use) or an output (e.g., air emissions). The assessment here proposed allows pursuing improvements of eco-efficiency by simultaneously decreasing environmental burdens and increasing benefits.

$$Max\ \beta_k \tag{1}$$

$$\sum_j \lambda_j y_{rj} \geq y_{rk} + \beta_k g_{y_r} \qquad\qquad r = 1, \dots, R \tag{1.1}$$

$$\sum_j \lambda_j h_{qj} \geq h_{qk} + \beta_k g_{h_q} \qquad\qquad q = 1, \dots, Q \tag{1.2}$$

$$\sum_j \lambda_j p_{lj} \leq p_{lk} - \beta_k g_{p_l} \qquad\qquad l = 1, \dots, L \tag{1.3}$$

$$\lambda_j \geq 0 \qquad\qquad j = 1, \dots, N$$

$$\beta_k \geq 0$$

In model (1), y_{rj} is the amount of economic benefit r generated by DMU j (with $r = 1, \dots, R$ and $j = 1, \dots, N$), h_{qj} corresponds to the environmental benefit q generated by DMU j (with $q = 1, \dots, Q$ and $j = 1, \dots, N$), and p_l is the environmental burden l generated by DMU j (with $l = 1, \dots, L$ and $j = 1, \dots, N$).

The indicators used for representing these dimensions in the assessment of mining firms are those presented in Table 1. The index k designates the DMU under assessment in each linear programming model. The directional vector $g = \left(g_{y_r}, g_{h_q}, -g_{p_l} \right)$ specifies the direction of projection to the frontier used to obtain the Eco-efficiency score. Positive values mean that the indicators should be increased, whereas negative values mean that the indicators should be reduced. The components of the directional vector specified in the empirical study correspond to the observed values of the indicators for the DMU k under assessment. This allows a radial interpretation of the objective function value obtained by the DDF model. The left-hand side of the constraints (1.1) to (1.3) describes the convex combination of the peers, corresponding to the point on the frontier against which DMU k is compared to estimate the Eco-efficiency level.

Variable β_k is a measure of Eco-efficiency. It corresponds to the rate by which the benefits and burdens of DMU k can be adjusted to achieve eco-efficient levels. When $\beta_k = 0$ the DMU can no longer proportionally enhance its Eco-efficiency indicators and thus it is classified as efficient. Values of β_k greater than zero correspond to inefficient DMUs. The Eco-efficiency score can be here interpreted as the maximum feasible radial adjustment to benefits and burdens leading to firms operation at the frontier of the production possibility set.

4 Application: Mining Companies Assessment

Our real world application explored a sample of 25 world-class mining companies that published their environmental and ecological outcomes in sustainability reports. The reference year of assessment is 2011. All the companies studied are affiliated to the Global Reporting Initiative (GRI) and 80% of them are members of the ICMM.

The dataset had variables with missing data (3% of all data) and values equal to zero (16% of all data). For the indicators to be maximized (y_r and h_q), the missing data were replaced by the lowest positive value observed in the corresponding variable. For the indicators to be minimized (p_k), the missing values were replaced by the highest value observed in the corresponding variable. These procedures guarantee that no DMU assessed will be unduly benefited for not having data available for some indicators. Zeros were always replaced by the lowest value of the corresponding variable, in order to improve the discrimination of the model.

Three alternative scenarios were exploded to investigate inefficiencies using different perspectives. These allow the identification of improvements aligned with specific managerial preferences. Scenario 1 focuses on improvements to the environmental dimension, using the directional vector $g_1 = \left(0, h_{qk}, -p_{lk} \right)$. Scenario 2 allows exploring the potential for reducing exclusively environmental burdens, using the directional vector $g_2 = (0, 0, -p_{lk})$. Scenario 3 explores potential enhancements of environmental benefits according to the directional vector $g_3 = \left(0, h_{qk}, 0 \right)$.

The results of the extended Eco-efficiency assessment revealed that 20 firms can be considered efficient in all scenarios explored. The score for the inefficient firms is reported in Table 2, alongside their ranking. The Eco-efficiency score $_k$ depends on the preferences specified in the directional vector regarding the performance improvements. For example, in the case of Sumitomo (U23), the radial improvement potential considering a proportional adjustment to all environmental indicators is 0.711 (see scenario 1). When each environmental category is explored in detail, we can conclude that the greatest potential for improvement lies on indicators representing environmental benefits, as the proportional improvement in scenario 3 (β_{U23} = 4.926) is higher than the proportional reduction to environmental burdens explored in scenario 2 (β_{U23} = 0.831). Regarding the ranking of inefficient DMUs, Table 2 shows that the ordering is not affected by the perspective of the assessment.

Table 2 Values of β_k for the extended Eco-efficiency assessment in 3 scenarios

	Scenario 1		Scenario 2		Scenario 3	
	β_k	Rank	β_k	Rank	β_k	Rank
Vale (U1)	0.108	1	0.195	1	0.242	1
Yamana (U15)	0.377	2	0.547	2	1.208	2
Gold Fields (U17)	0.411	3	0.583	3	1.395	3
Hydro (U8)	0.637	4	0.779	4	3.512	4
Sumitomo (U23)	0.711	5	0.831	5	4.926	5

Model (1) also enables obtaining proportional improvement targets for each indicator, calculated using the Eco-efficiency score β_k. Table 3 presents the results obtained for Gold Fields (U17) to illustrate the interpretation of these results.

Table 3 Targets for Gold Field (U17)

	Observed	Scenario 1 Target	Scenario 2 Target	Scenario 3 Target
Enconomic Benefits				
Value-added (y_1)	3688.000	–	–	–
Environmental Benefits				
Renewable energy use (h_1)	0. 000	0.000	–	0.000
Recycled materials use (h_2)	*	*	*	*
Water recycled or reused (h_3)	33.450	47.198	–	80.11275
Conservation areas suported (h_4)	0. 000	0.000	–	0.000
Investiments on environment (h_5)	677. 000	955.247	–	1621.415
Environmental Burdens				
Non-renewable energy use (p_1)	5469784. 000	3221702.776	2280899.928	–
Withdraw water use (p_2)	78236. 000	46081.004	32624.412	–
Waste produced (p_3)	15000000. 000	8835000.000	6255000.000	–
Air emissions (p_4)	5.298	3.121	2.209	–
Volume of significant spills (p_5)	47000. 000	27683.000	19599.000	–
Environmental Fines (P_6)	*	*	*	*

* missing value

Considering scenario 1, Gold Fields (U17) obtained an Eco-efficiency score equal to 0.411. This means that this firm should increase environmental benefits proportionally by the factor $(1 + \beta_k)$ and reduce the burdens by the factor $(1 - \beta_k)$. For example, the indicator of withdraw water use (p_2) could reduce from 78236.000 to 46081.004 (i.e. $78236 \times (1-0.411) = 46081.004$) and investments on environment (h_5) could increase from 677.000 million USD to 955.247 million USD (i.e. 677.000 $\times (1+0.411) = 955.247$). When exploring the potential for environmental benefits in isolation (scenario 3), without seeking changes to indicators from other categories, the target for investments on environment (h_5) is more demanding, reaching 1621415 million USD (i.e. 677.000 $\times (1+1.395) = 1621.415$). A similar interpretation can be done for scenario 2, as the potential for reductions to environmental burdens in isolation is higher than when the firms have to focus on simultaneous improvements to environmental benefits and reductions to burdens.

5   Conclusions

In this study, we proposed an enhanced Eco-efficiency assessment model, aligned with the definition of the WBCSD. It advances the traditional measure of Eco-efficiency from the ratio of economic benefits to environmental burdens in order to also include environmental benefits related to companies' activities. Incorporating these new features in the Eco-efficiency analysis can help the design of policies to improve economic performance alongside avoiding undesirable impacts on the ecosystem or the exhaustion of natural resources. This can be achieved by either promoting the use of environmental friendly resources or investing in conservation plans.

By widening the focus of the Eco-efficiency evaluation, we believe this study succeeded in proposing a new framework to assess mining companies. In this sector, the efforts to reduce environmental impacts are particularly important. With the use of the model proposed in this paper, the companies that invest on environmental benefits can be credited with higher Eco-efficiency scores. Another noteworthy feature of this study is the possibility of setting reduction goals for air emissions, waste generation and materials consumption, alongside improvements to economic and environmental indicators that are associated with benefits both for the firms and the planet. Furthermore, customized directional vectors allowed exploring alternative scenarios for improvements in Eco-efficiency, aligned with specific managerial preferences.

We foresee as future research the assessment of Eco-efficiency over time, so that the evolution of environmental practices can be tracked systematically.

Acknowledgments. Thanks are due to CAPES for funding this work through the Program Science without Borders. (BEX 19131127). Thanks are also due to University of Para State (UEPA).

References

Chambers, R.G., Chung, Y., Färe, R.: Benefit and Distance Functions. Journal of Economic Theory 70(2), 407–419 (1996)

Charnes, A., Cooper, W.W., Rhodes, E.: Measuring the efficiency of decision making units. European Journal of Operational Research 2(6), 429–444 (1978)

Choucri, N.: Globalisation of eco-efficiency: GSSD on the WWW. In: UNEP Industry and Environment, pp. 45–49. United Nations, New York (1995)

Cramer, J.: How can we substantially increase eco-efficiency? Industry and Environment 20(3), 58–62 (1997). http://www.scopus.com/inward/record.url?eid=2-s2.0-6044267126&partnerID=tZOtx3y1

Esty, D.C., Porter, M.E.: Industrial Ecology and Competitiveness. Journal of Industrial Ecology 2, 35–43 (1998). http://digitalcommons.law.yale.edu/fss_papers/444

Färe, R., Grosskopf, S., Pasurka, C.A.: Potential gains from trading bad outputs: The case of U.S. electric power plants. Resource and Energy Economics 36(1), 99–112 (2014)

Glauser, M., Müller, P.: Eco-efficiency: A prerequisite for future success. Chimia 51(5), 201–206 (1997)

GRI, Global Reporting Initiative. Global Reporting Initiative, p.1 (2013). https://www.globalreporting.org/Pages/GRIOrganizationsSearchPage.aspx

ISO, 2003. ISO 14000 - Environmental management, pp.1–19

Kortelainen, M., Kuosmanen, T.: Eco-efficiency analysis of consumer durables using absolute shadow prices. Journal of Productivity Analysis 28(1–2), 57–69 (2007)

Picazo-Tadeo, A.J., Prior, D.: Environmental externalities and efficiency measurement. Journal of Environmental Management 90(11), 3332–3339 (2009)

Schaltegger, S., Sturm, A.: Öologische Rationalität (German/in English: Environmental rationality). Die Unternehmung 4, 117–131 (1990)

Schmidheiny, S.: Changing course: A global perspective on development and the environment, 1st edn. The MIT Press, Cambridge (1992)

WBCSD, Eco-Efficiency: creating more value with less impact 1st edn. WBCSD, Geneva (2000)

Zhang, B., et al.: Eco-efficiency analysis of industrial system in China: A data envelopment analysis approach. Ecological Economics 68(1–2), 306–316 (2008)

Appendix

Table A.1 Companies dataset of benefits

Companies	y_1 mil. USD	h_1 GJ	h_2 mil. m^3	h_3 mil. m^3	h_4 ha	h_5 mil. USD
Vale (U1)	85.000	24000.000	0.000	953.000	2250.000	525.000
Alcoa (U2)	545.008	0.000	59.890	0.000	27816.460	8.132
Anglo American Ni (U3)	398.000	1106182.990	0.000	2823.200	11000.000	69.222
Rio Norte (U4)	0.778	1.000	2378.700	39.864	429600.000	9.280
Sama (U5)	94.609	0.000	62.760	0.000	2500.000	2.171
Samarco (U6)	1767.741	0.000	663.360	158.547	2228.490	94.300
Votorantim (U7)	5383.523	0.000	80.260	227.220	0.000	150.130
Alunorte Hydro (U8)	70.588	0.000	0.320	11.613	4600.000	10.691
Kinkross (U9)	1382.100	0.000	104.600	321.000	5661.000	15.000
Usiminas (U10)	1493.254	178680436.000	1.670	44.603	1250.000	92.838
Rio Tinto (U11)	38193.000	3000000.000	368.400	327.600	348500.000	21.168
Barrick (U12)	13000.000	38700000.000	4.332	24.471	118400.000	100.000
BHP (U13)	72299.000	30000000.000	1.181	79.640	25100.000	7.820
Glencore (U14)	0.000	16400000.000	0.074	240.000	289.000	3015.000
Yamana (U15)	7.120	6738121.137	0.000	3012.874	32700.000	9.328
Nipon (U16)	221000.000	3600.000	0.026	115.452	870.940	0.000
Gold Field (U17)	3688.000	0.000	*	33.453	0.000	677.000
MitiSubish Materials (U18)	22423.700	0.000	19.100	406.636	28600.000	7780.862
Gold Corp (U19)	2626.000	0.000	0.000	107.500	343100.000	107.500
Teck (U20)	0.500	13394000.000	0.000	200.839	28025.000	200.839
ARM (U21)	1007.248	0.000	0.000	0.000	11891.000	2.171
Coldeco (U22)	23083.000	0.000	0.779	490.165	2.651	490.165
Sumitomo (U23)	1597.696	0.000	0.160	0.000	151.000	2.171
De Beers (U24)	6400.000	0.000	0.000	30.110	195640.000	30.110

*missing value

Table A.2 Companies dataset of environmental burdens

Companies	p_1 GJ	p_2 mil. m^3	p_3 tons	p_4 tons	p_5 m^3	p_6 mil. USD
Vale (U1)	179700.000	420.700	686.000	52190000.000	350.500	3019.000
Alcoa (U2)	55328005.000	0.205	80821.000	2886227.000	53.000	0.000
Anglo American Ni (U3)	4786.000	3529.000	3729.000	2070.930	0.727	*
Rio Norte (U4)	0.778	18.163	2587.900	320777.000	18.163	0.000
Sama (U5)	948.714	1.694	125.080	38206.000	0.200	0.000
Samarco (U6)	14208818.400	2083.664	19138.120	783350.300	0.000	1.770
Votorantim (U7)	294281.974	227.221	4270.000	26000000.000	0.000	0.109
Alunorte Hydro (U8)	170190000.000	34.780	8178.460	929110.000	0.000	5.375
Kinkross (U9)	14191000.000	104.528	222.553	1.220	0.000	2700.000
Usiminas (U10)	179976814.000	181.608	17.633	12635316.000	0.000	0.505
Rio Tinto (U11)	516000000.000	465.000	1166.600	631170.124	3.000	80150.000
Barrick (U12)	54408761.000	57.100	45.548	50307341.005	45.548	764800.000
BHP (U13)	286180000.000	181.000	35.367	21000.000	0.000	2454.000
Glencore (U14)	164000000.000	500.000	500.000	130698.000	570.000	210000.000
Yamana (U15)	19818003.345	3766.092	3766.092	229365.357	*	29193.850
Nipon (U16)	16627000.000	21.453	21.453	0.005	0.000	0.000
Gold Field (U17)	5469784.000	78236.000	15000000.000	5.298	47000	*
MitiSubish Mat. (U18)	3923600.000	14.620	18557.000	24.290	0.000	0.000
Gold Corp (U19)	15400000.000	144.800	490696.000	1.415	0.000	0.000

Table A.2 (*continued*)

Teck (U20)	30849.000	118.955	0.767	304.610	0.206	0.000
ARM (U21)	9179218.800	15091.358	0.444	2396.000	0.000	0.000
Coldeco (U22)	764.125	174.891	253.643	5592875.158	30.000	216.554
Sumitomo (U23)	31344106.000	31.500	5.415	4940628.000	*	*
De Beers (U24)	11590.000	7.110	1602.000	1450000.000	*	102600.000
Anglo Platinum (U25)	25170.000	18.980	1566.070	12000770.000	0.900	2000.000

*missing value

Exploring a Column Generation Approach for a Routing Problem with Sequential Packing Constraints

Telmo Pinto, Cláudio Alves, and José Valério de Carvalho

Centro ALGORITMI, Escola de Engenharia, Universidade do Minho, Portugal

Abstract. In this work we propose a computational study of a column generation based heuristic prototype for the vehicle routing problem with two-dimensional loading constraints. This prototype was recently proposed in literature (Pinto et al., 2013) and it relies in a column generation algorithm whose subproblem is relaxed. After solving the subproblem, the feasibility of the routes is verified using lower bounds and the classical bottom-left heuristic, enhanced with sequential constraints. For the infeasible routes, a simple route shortening process is applied, and the feasibility is tested again. The effectiveness of our approach is evaluated using benchmark instances from the literature.

1   Introduction

One important variant of the Vehicle Routing Problem (VRP) is the Capacitated Vehicle Routing Problem (CVRP). In the CVRP, a value (usually a weight or volume) is associated to the demand of the customers and thus each vehicle can only carry items which do not exceed its available capacity. However, in real-world context, taking into account just the capacity of the vehicle can lead to situations where this capacity is not violated, but items do not fit all in the loading area. For instance, if the demands are composed of bulky but not heavy items, and if only the weight capacity of the vehicle is considered, the items may not fit into the vehicles even if the weight limit is satisfied. For this purpose, loading constraints should be considered, since the distribution of the items along the loading area has a direct impact in the loading operations. These problems are known in the literature as CVRP with loading constraints (L-CVRP) and integrate the resolution of the CVRP with the definition of the loading positions of the items. This latter issue is typically treated with the resolution of a Bin Packing Problem (BPP) or a Strip Packing Problem (SPP) in either two- (2L-CVRP) or three-dimensions (3L-CVRP). The complexity of these problems is NP-hard since L-CVRP is a generalization of the CVRP.

The relative positions of the items in vehicles play an important role in the problem. These positions can be defined taking into account the sequence in which customers are served, in order to avoid rearranging items after unloading the demand of a certain customer. Specifically, the sequential constraints impose that unloading an item to a customer precludes movements in other items. Consequently, the items must be loaded in the vehicle in the opposite order in which they are unloaded, following a LIFO policy (*Last-In, First Out*). The inclusion of sequential constraints in the L-CVRP is known as sequential L-CVRP. Otherwise, the problem is denominated by unrestricted L-CVRP.

Due to the complexity of the L-CVRP, the vast majority of the approaches are based on heuristics. To the best of our knowledge, only one exact approach for the 2L-CVRP is presented in literature, and it is due to Iori et al. (2007). This approach relies on a branch-and-cut method. In each iteration, a branch-and-bound algorithm is applied to verify the feasibility of the loading. In this algorithm, and from an empty packing solution, one node is generated for each item which is placed at the bottom-left corner of the surface. For each generated node, a new node is created in the downward level for each item that has not been placed and for each feasible position. In the overall approach, some heuristic separation methods are used to search for violated inequalities. This approach presents good results, solving instances with up to 35 customers and 100 items to the optimality.

The first heuristic approach for the 2L-CVRP is due to Gendreau et al. (2008) who performed a tabu search algorithm. In order to verify the loading feasibility, the customers are sorted by the reverse order of visit and for each customer the items are sorted by decreasing order of width. The first item is placed at the origin (0, 0) and the following items are placed in the position that maximizes the touching perimeter (Lodi et al. 1999. Movements to neighbor solutions can lead to routes which violate the weight capacity or height of the vehicle. If these situations happen, these moves are penalized in the objective function.

In 2009, a guided tabu search algorithm was proposed by Zachariadis et al. (2009) for the 2L-CVRP. The objective function incorporated in the Tabu Search algorithm is modified in order to increase diversification, guiding the search to unexplored solutions. To verify the loading feasibility of the problem the authors presented a heuristic bundle composed by five packing heuristics which are successively applied until a valid solution is found.

Based on the work of Zachariadis et al. (2009), Leung et al. (2011) proposed an Extended Guided Tabu Search for the 2L-CVRP. This approach aims to apply a guided local search extension based on the fact that if one move can lead to a new best solution, then the penalty associated to that move must be suppressed.

Fuellerer et al. (2009) proposed the first work for the 2L-CVRP with variable orientation on items. This approach is an Ant Colony Optimization algorithm. During the local search procedure, if any solution requires more vehicles than those that compose the fleet, the objective function will be penalized and the pheromone information is not updated.

This paper is organized as follows. In Section 2 we describe the sequential 2L-CVRP. In Section 3, our approach is presented. Some computational results are presented in Section 4. Finally, some conclusions are drawn in Section 5.

2   Problem Description

Following the notation introduced by Gendreau et al. (2008), the Vehicle Routing Problem with Two-dimensional Loading Constraints (2L-CVRP) can be formulated in a complete undirected graph $G=(V,E)$ where $V=\{0,1,...,n\}$ is the set of $n+1$ vertices corresponding to the depot (vertex 0) and to the n customers, and E is the edge set connecting all pair of vertices such that $E=\{(i,j): i,j \in V, i \neq j\}$. The cost of traveling along an edge $(i,j) \in E$ is represented by c_{ij}. There is a homogeneous fleet composed by K vehicles with a two-dimensional rectangular loading surface with height H and width W. The demand of each customer i is composed by m_i two-dimensional rectangular items.

The 2L-CVRP consists in finding a set of optimal routes S starting and ending at the depot, satisfying the following constraints. The number of routes of set S cannot be greater than the fleet size K (C1) and all customers must be served in a single visit (C2). The items may have a fixed orientation (C3) and must be completely within the surface, must not overlap, and the edges of the items must be parallel to the surface edges (C4). Moreover, unloading an item to a customer does not require moving the other items for customers which are served later. Therefore, the items must be placed inside the vehicle according to *Last-In First Out* policy. These constraints are known as sequential constraints (C5).

3   Column Generation Approach

This paper relies on the prototype of the column generation approach described in Pinto et al. (2013). Since no computational experiments were conducted, this paper aims to contribute with a computational study of the proposed prototype with benchmark instances. The proposed prototype works as follows.

Generally speaking, we solve a sequence of linear relaxations of a restricted master problem, while the pricing subproblem is relaxed and solved using heuristic procedures. In order to generate attractive columns we propose a relaxed subproblem, where no loading constraints are considered. However, a new constraint is added to the subproblem requiring that the sum of the area of the items inside the vehicle should be less or equal than the area of the vehicle. Additionally, after deriving an attractive column, lower bounds of Martello and Vigo (1998) are applied. These bounds were used by Iori et al. (2007) in order to evaluate the possible infeasibility of the route. If the route is clearly infeasible, one customer is dropped from the route and the process is repeated, taking into account that the reduced cost of the column has to remain negative. If it is not possible to prove the infeasibility of the route, a valid packing can be derived using a bottom-left heuristic with sequential constraints. The bottom-left heuristic was used by some authors aiming to derive a valid packing (Iori et al., 2007; Zachariadis et al., 2009; Leung et al., 2011). However, if this heuristic is not able to produce a feasible packing, a customer is dropped again from the route, only if the reduced cost remains negative. The bottom-left heuristic with sequential constraints is repeated for the resulting route. In order to avoid a significant impact in the reduced cost,

the selected customer to be dropped is the one whose associated dual variable has
the lowest absolute value.

If a valid route is found, the corresponding column is added to the restricted
master problem. Otherwise, the algorithm stops. At this stage, the restricted master
problem is solved up to integrality with the set of columns generated in the previous steps. This procedure aims to both produce an integer feasible solution for the
2L-CVRP, and also to evaluate the size of the optimality gap whenever the optimal solution of the linear relaxation of the master problem is obtained. Figure 1
summarizes the described approach.

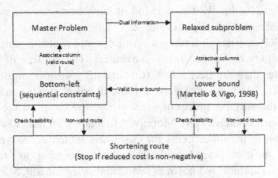

Fig. 1 Overall approach.

One can apply the Dantzig-Wolfe decomposition to a standard compact formulation of the 2L-CVRP, as in (Pinto et al., 2013). As a result, we get a master
problem with a set-partitioning structure. Let Ω be the set of all feasible routes.
Each route r with cost c_r is defined by a vector $(a_{1r}, a_{2r}, ..., a_{n,r})^T$ where a_{ir} is the
coefficient which indicates if customer i is visited in route r for $i \in \{1,...,n\}$. One
decision variable y_r is associated to each route $r \in \Omega$, taking value one if the route
is included in the restricted master problem, and zero otherwise. Each feasible
route is in fact a column of the restricted master problem. The same subset of customers can be visited in more than one route $r \in \Omega$, and thus more than one column can have the same vectors. Nevertheless, they are in fact distinct routes, with
different ways of visiting customers and thereafter the cost of the column c_r is
different.

The restricted master problem can be formulated as follows.

$$\text{Minimize} \quad \sum_{r \in \Omega} c_r y_r \tag{1}$$

$$\text{Subject to} \quad \sum_{r \in \Omega} a_{ir} y_r = 1, \forall i \in \{1,...,n\} \tag{2}$$

$$\sum_{r \in \Omega} y_r \leq K \tag{3}$$

$$y_r \in \{0,1\}, \forall r \in \Omega \tag{4}$$

The subproblem can be formulated as an Elementary Shortest Path Problem with Capacity Constraints (ESPPCC). Let π_i ($i \in V\backslash\{0\}$) be the dual variables of the restricted master problem associated to the constraints set (2). Let μ be the dual variable associated to constraint (3). Additionally, let x_{ij}, $\forall$ $(i,j) \in E$ be the decision variables which take value one if the arc $(i,j) \in E$ is included in the optimal solution, and take value 0 otherwise.

As alluded above, the loading constraints are relaxed in the subproblem, and a new constraint is considered instead. This constraint requires that the sum of the area of the items of each visited customer j ($area_j$) must be less or equal than the area of the vehicle ($areaV$).

Thus, the subproblem can be formulated as follows.

$$\text{Minimize} \qquad \sum_{(i,j)\in E} c'_{ij} x_{ij} \qquad (5)$$

$$\text{Subject to:} \qquad \sum_{i\in V\backslash\{0\}} x_{0i} = \sum_{i\in V\backslash\{0\}} x_{i0} = 1 \qquad (6)$$

$$\sum_{i\in V} x_{ij} - \sum_{i\in V} x_{ji} = 0, \forall j \in V \backslash \{0\} \qquad (7)$$

$$\sum_{i\in S} \sum_{j\in S} x_{ij} \leq |S| - 1, \forall S \subseteq V \backslash \{0\}, S \neq \varnothing \qquad (8)$$

$$\sum_{j\in V} area_j \sum_{(i,j)\in E} x_{ij} \leq areaV \qquad (9)$$

$$x_{ij} \in \{0,1\}, \forall (i,j) \in E \qquad (10)$$

$$\text{with} \qquad c'_{ij} = \begin{cases} c_{ij} - \pi_i, \forall (i,j) \in E, i \neq 0 \\ c_{ij} - \mu, \forall (0,j) \in E \end{cases}$$

4 Computational Results

Our approach was tested using benchmark instances (Iori et al., 2007; Iori, 2004, Gendreau et al., 2008). Only instances with up to 50 customers were considered. These instances, are divided in 5 classes. For the first class, each customer demands a single item with both width and height size equal to one (pure CRVP instances). For the following classes, the number of required items for each customer is obtained by a uniform distribution in the interval $[1, n]$ where n is the number of the class.

The restricted master problem is initialized with single-customer routes, with routes with a subset of customers with less than 90% of the cargo, and with 100 columns which randomly selects p customers with $p=n/K$. These two last strategies use the nearest neighbor heuristic to determine the sequence. All these columns are validated using lower bounds and the bottom-left heuristic with sequential constraints. In order to solve the subproblem, we use the dynamic programming approach algorithm of Righini and Salani (2008) which performed good results in other works (Salani and Vacca, 2011; Vacca et al., 2013).

In all the results for instances of Class 1, our column generation approach was able to find the optimal of the linear relaxation (column z_{RL}). These results are presented in Table 1. In 6 over the 19 instances for this class, the linear relaxation is also the optimal integer solution. For the remaining instances, an upper bound is provided, in column z_{PI}, with exception for two instances which it was not possible to achieve this bound. This value is obtained by solving the restricted master problem up to optimality and integrality as referred above.

The number of generated columns, the number of customers and the number of vehicles are presented in columns *Cols*, n and K respectively. The total execution time in seconds is presented in column *Time* and finally the relative integrality gap $((z_{PI^-} z_{RL})/ z_{PI})*100)$ is presented in column *GAP*. The best achieved value in literature is provided in column z_{OPT}.

Table 1 Results for Class 1 instances

Instance	n	K	z_{RL}	Cols	z_{PI}	Time	GAP	z_{OPT}
1. E016-03m	15	3	215	203	215	1,27	0	273
2. E016-05m	15	5	214	194	214	0,86	0	329
3. E021-04m	20	4	250	269	250	3,03	0	351
4. E021-06m	20	6	255	231	255	1,8	0	423
5. E022-04g	21	4	283,71	247	322	2,75	11,89	367
6. E022-06m	21	6	277,8	304	284	3,25	2,18	488
7. E023-03g	22	3	487,5	266	514	2,91	5,16	558
8. E023-05s	22	5	490,91	245	566	2,41	13,27	657
9. E026-08m	25	8	310,4	331	397	4,51	21,81	609
10.E030-03g	29	3	383	414	383	9,26	0	524
11. E030-04s	29	4	400,88	455	418	9,79	4,1	500
12. E031-09h	30	9	310,28	449	334	9,39	7,1	-
13. E033-03n	32	3	1805,08	506	-	15,7	-	1991
14. E033-04g	32	4	430	588	430	31,24	0	-
15. E033-05s	32	5	463,97	415	639	20,55	27,39	907
16. E036-11h	35	11	343,36	603	408	19,99	15,84	682
17. E041-14h	40	14	355,94	742	462	37,39	22,96	-
18. E045-04f	44	4	651,04	621	694	65,57	6,19	-
19. E051-05e	50	5	419,28	895	-	89,29	-	-

In instances of classes 2 to 5, this approach was not able to produce the optimal value of the linear relaxation (the best achieved value is presented in column *Best*. However, in order to evaluate the quality of the overall approach, we present in Table 2 the number of failures when checking the feasibility of a route. In this sense, column *FL* presents the number of times that the lower bound is greater than 1 and *FB* presents the number of times in which the bottom-left with sequential constraints was not able to produce a feasible loading. Again, an upper bound is provided (column z_{PI}) by imposing the integrality condition. Each instance (column *I*) is numbered, according to Table 1, and column *C* indicates the class.

Table 2 Results for Class 2 to Class 5

I	C	Best	Cols	FL	FB	z_{PI}	T	I	C	Best	Cols	FL	FB	z_{PI}	T
1	2	426,33	124	0	22	-	0,3	10	4	777,46	164	0	118	820	1,3
1	3	-	123	0	21	-	0,2	10	5	706,27	191	0	179	820	3,3
1	4	425,00	128	0	49	425	0,3	11	2	783,92	153	0	88	820	1,0
1	5	318,97	134	0	53	326	0,4	11	3	761,83	173	0	170	820	1,8
2	2	364,89	128	0	26	392	0,2	11	4	776,39	159	2	90	820	1,0
2	3	346,65	129	0	20	367	0,1	11	5	764,78	153	0	80	820	1,1
2	4	366,33	127	0	25	399	0,2	12	2	692,10	156	2	59	832	1,0
2	5	286,17	141	0	66	316	0,6	12	3	590,53	173	0	154	695	2,0
3	2	497,13	131	0	18	526	0,2	12	4	641,24	165	0	111	673	1,4
3	3	525,14	133	0	31	-	0,6	12	5	588,29	178	0	126	663	2,0
3	4	471,06	137	0	56	605	0,6	13	2	3920,08	160	2	114	5994	2,5
3	5	410,44	146	0	100	556	0,8	13	3	3991,98	153	0	94	5994	2,2
4	2	470,73	144	0	64	486	0,5	13	4	3650,71	167	0	154	5994	4,6
4	3	475,33	147	0	79	533	1,1	13	5	3515,64	161	0	109	5994	2,4
4	4	525,20	136	0	34	595	0,3	14	2	1196,00	154	5	71	1196	1,6
4	5	425,17	148	0	83	462	0,7	14	3	1115,19	169	0	173	1196	4,3
5	2	688,10	134	0	43	-	1,0	14	4	1121,62	184	0	249	1196	5,7
5	3	537,01	148	0	101	-	1,6	14	5	1108,40	182	0	258	1196	6,3
5	4	608,86	140	0	82	-	1,4	15	2	1196,00	151	0	52	1196	1,3
5	5	510,44	141	0	71	571	0,7	15	3	1196,00	150	0	56	1196	1,0
6	2	525,24	145	4	70	630	0,7	15	4	1196,00	161	0	103	1196	1,6
6	3	551,96	142	0	47	615	0,4	15	5	1196,00	160	0	113	1196	2,0
6	4	565,60	136	0	23	621	0,2	16	2	722,52	184	0	151	729	2,2
6	5	476,50	145	0	65	585	0,6	16	3	703,08	182	5	128	827	2,0
7	2	941,19	141	0	70	1287	0,7	16	4	855,58	161	0	55	1038	0,9
7	3	907,65	148	0	105	1043	0,9	16	5	609,80	207	0	233	772	4,6
7	4	1030,65	133	0	41	1287	0,5	17	2	946,70	185	0	140	1118	4,0
7	5	734,50	170	0	162	933	1,5	17	3	767,01	206	0	232	819	6,0
8	2	899,38	151	0	117	1287	0,9	17	4	824,27	220	0	307	940	6,4
8	3	907,41	148	0	99	1287	0,9	17	5	644,76	246	0	398	806	10,8
8	4	1085,97	132	0	41	1287	0,5	18	2	1559,00	169	0	93	1559	3,8
8	5	694,55	173	0	261	820	2,4	18	3	1357,91	184	0	144	1559	5,8
9	2	737,29	145	0	38	757	0,5	18	4	1559,00	166	0	87	1559	3,0
9	3	782,07	142	0	26	864	0,4	18	5	1306,75	202	0	313	1559	16,7
9	4	624,68	175	2	152	670	1,3	19	2	1126,24	207	14	223	1625	11,2
9	5	617,96	165	0	123	690	1,4	19	3	1119,69	225	0	322	1625	17,3
10	2	782,39	154	0	84	820	1,3	19	4	1166,42	204	0	225	1625	8,9
10	3	751,43	172	0	161	820	2,3	19	5	869,54	263	0	462	1614	24,19

The results presented in Table 2 show that within a short period of time is possible to insert in the restricted master problem more than one hundred of columns which are in fact valid solutions of 2L-CVRP, even when the validation is performed with simple methods as the bottom-left heuristic with sequential constraints. By other hand, the number of failures when trying to derive a feasible solution using this heuristic is high.

5 Conclusion

In this paper, we explored the application of a column generation based approach to a vehicle routing problem with two-dimensional and sequential packing constraints. From a prototype presented in literature, we provided a computational study on benchmark instances which aims to contribute with experimental results obtained in the resolution of such problems with column generation support. We showed that in some cases the approach appears to be promising concerning the overall execution time. On the other hand, the results considering the class instances other than class 1 show that the validation of the routes with a simple bottom-left with sequential constraints presents obvious limitations. This aspect in particular will be considered in future developments.

Acknowledgments. This work was supported by FEDER funding through the Programa Operacional Factores de Competitividade - COMPETE and by national funding through the Portuguese Science and Technology Foundation (FCT) in the scope of the project PTDC/EGE-GES/116676/2010 (Ref. COMPETE: FCOMP-01-0124-FEDER-020430), by FCT within the project scope: UID/CEC/00319/2013, and by FCT through the grant SFRH/BD/73584/ 2010 for Telmo Pinto (funded by QREN - POPH - Typology 4.1 - co-funded by MEC National Funding and the European Social Fund), and by FEDER funds through the Competitiveness Factors Operational Programme - COMPETE.

References

Fuellerer, G., Doerner, K., Hartl, R., Iori, M.: Ant colony optimization for the two-dimensional loading vehicle routing problem. Computers & Operations Research 36(3), 655–673 (2009)

Gendreau, M., Iori, M., Laporte, G., Martello, S.: A Tabu search heuristic for the vehicle routing problem with two-dimensional loading constraints. Networks 51(1), 4–18 (2008)

Iori, M.: Metaheuristic algorithms for combinatorial optimization problems. PhD thesis, DEIS, University of Bologna (2004)

Iori, M., Salazar-Gonzalez, J.-J., Vigo, D.: An exact approach for the vehicle routing problem with twodimensional loading constraints. Transportation Science 41(2), 253–264 (2007)

Leung, S.C., Zhou, X., Zhang, D., Zheng, J.: Extended guided tabu search and a new packing algorithm for the two-dimensional loading vehicle routing problem. Computers & Operations Research 38(1), 205–215 (2011)

Lodi, A., Martello, S., Vigo, D.: Heuristic and metaheuristic approaches for a class of two-dimensional bin packing problems. INFORMS Journal on Computing 11(4), 345–357 (1999)

Martello, S., Vigo, D.: Exact solution of the two-dimensional finite bin packing problem. Management science 44(3), 388–399 (1998)

Pinto, T., Alves, C., de Carvalho, J.V.: Column generation based heuristic for a vehicle routing problem with 2-dimensional loading constraints: a prototype. In: XI Congresso Galego de Estatística e Investigación de Operacións, Spain (2013)

Righini, G., Salani, M.: New dynamic programming algorithms for the resource constrained elementary shortest path problem. Networks 51(3), 155–170 (2008)

Salani, M., Vacca, I.: Branch and price for the vehicle routing problem with discrete split deliveries and time windows. European Journal of Operational Research 213(3), 470–477 (2011)

Vacca, I., Salani, M., Bierlaire, M.: An exact algorithm for the integrated planning of berth allocation and quay crane assignment. Transportation Science 47(2), 148–161 (2013)

Zachariadis, E.E., Tarantilis, C.D., Kiranoudis, C.T.: A guided tabu search for the vehicle routing problem with two-dimensional loading constraints. European Journal of Operational Research 195(3), 729–743 (2009)

Design of Retail Backroom Storage: A Research Opportunity?

Maria Pires[1], Pedro Amorim[1], Jorge Liz[2], and Joaquim Pratas[2]

[1] INESC TEC and Faculty of Engineering, University of Porto, Porto, Portugal
[2] SONAE MC, Porto, Portugal

Abstract. The design of retail backroom storage has a great impact on in-store operations, customer service levels and store life-cycle costs. Moreover, backroom storage in modern retail stores is crucial to several functions, such as acting as a buffer against strong demand lifts yielded by an increasing promotional activity, seasonal peak demand and e-commerce activities. Despite having similar functions to a distribution center, backroom storage facilities have particularities that deserve a distinct analysis. In this paper we aim to draw attention to the lack of research about this topic.

1   Introduction

Warehouses are a key part of modern supply chains and play a vital role in the success or failure of business today [1]. Additionally, the large scale retail business is one of the most important supply chain stages both in terms of revenue per year and number of actors and entities involved [2]. It is estimated that the operating costs and capital invested in warehouses represent 22% of logistics costs in USA and 25% in Europe, and that in-store logistics are the most costly part of the retailer's supply chain [3][4]. The increased competition in the retail industry has required continual improvements in design and operation of the supply chain, which also requires a better performance by the warehouses [5].

Backroom storage, highlighted in yellow in Figure 1, is essential in retail stores since the replenishment orders for a given item that arrive at a retail store, coming directly from suppliers or from distribution centers (DCs), may not fit on the allocated shelf space, making this area indispensable [6][7]. Moreover, nowadays, backroom storage in retail food stores is becoming more vital to act as a buffer against strong demand lifts yielded by an ever increasing promotional activity, seasonal peak demand for particular categories of products and on weekends, as well as to accommodate other activities such as e-commerce [5][8]. Promotional activities are a key aspect to retail stores, being a competitive factor, which has a strong influence in the customer loyalty [9]. Decisions regarding promotional activities and modification in planograms are often centrally defined and they strongly affect store operations since they require adjustments in the sales area,

© Springer International Publishing Switzerland 2015
A.P.F.D. Barbosa Póvoa and J.L. de Miranda (eds.), *Operations Research and Big Data,*
Studies in Big Data 15, DOI: 10.1007/978-3-319-24154-8_20

which is a very time and staff consuming process, and large quantities of merchandise in the backroom [8][10][11].

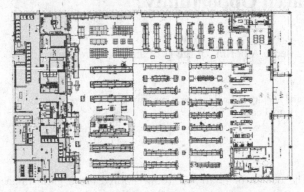

Fig. 1 Example of a store layout, with the backroom area in yellow

Despite having similar functions to a distribution center, backroom storage facilities have particularities that deserve a distinct analysis and which we aim to draw attention to. Operations on a retail store level are more complex and unorganized than in DCs [6]. This is largely explained by in-store logistics that includes frequent promotional campaigns, handling the flow of products between shelves, temporary storage areas, promotional areas and backroom areas [8][12]. Further, on a store level, order packaging units are smaller and more heterogeneous and customers exhibit a higher variability of demand. Moreover, stores stock a high range of products with specific characteristics (such as perishability, sensitivity to temperature and high shelf turnover), and deal with problems such as insufficient and busy staff, receiving errors and inventory shrinkage from theft, spoilage or damage [11][13].

Backroom storage design involves several research areas and it has been overlooked in the literature. The research areas related with the backroom design are very disperse and concern the conventional grocery retailing, warehouse design and operations, store operations, logistics and the facility layout problem. Despite the work undertaken in these distinct areas, a general framework linking these subjects in light of the backroom particularities is missing. For example, the models for designing conventional warehouses and DCs do not adjust to the necessities of backroom storage facilities, such as accommodating e-commerce operations, being robust against an intense promotional activity stress, and coping with in-store operations [14].

The remainder of this paper is organized as follows. The next section describes the research approach and methodology. Section 3 explains the particularities of backroom storage and describes briefly the operations within retail stores. Section 4 addresses the backroom design problem. Finally, Section 5 concludes the research paper and proposes future works.

2   Research Approach

The aim of this paper is to introduce the importance of backrooms in retail supply chains as well as their role at in-store operations and performance. A literature review was undertaken, from May to September of 2014, searching a range of electronic databases, including Science Direct, Google Scholar, Springer and Scopus. These databases were searched using combinations of relevant keywords, such as "backroom", "back store", "grocery retail", "design", "dimensioning" and "operations". Relevant papers were then selected based on the abstracts analysis. From this initial selection, the search was extended by accessing relevant books and cited papers. A total of 107 papers were retrieved that met at least one of the search terms. Selected papers date between 1984 and 2014. The literature was then classified into two groups: those that addressed the warehouse design and those that focused on grocery retail operations. It should be referred that no relevant literature existed in design of retail backroom storage areas *per se*.

In parallel with literature review, retail stores of a Portuguese retail company were analyzed. Data was obtained through observation and interviews with the store managers and personnel, when visiting a representative group of stores during January of 2015, as well as data extracted from the company historical data. In this company, grocery stores are divided in three segments: hypermarket, supermarkets and convenience stores. The first type of store can be characterized by a total of 40 stores with average sales area of approximately 7380 m^2, the second type has the greatest number of stores (128) and an average sales area of approximately 2093 m^2 and the last, with a total of 42 stores, has an average sales area of approximately 1060 m^2. For all types of stores the backroom storage areas occupies nearly 36% of the total area of the store.

3   Backroom Particularities

In most retail stores, inventory is held in two locations: retail shelves, in the sales area, and in the backroom, also called back store area. Products are stored in the backroom for many reasons but one main factor is the limited shelf space that makes it often impossible to fit a complete replenishment order on the allocated shelf space. By storing some inventory in the backroom, shelf space is freed for displaying a wider product assortment, potentially increasing sales [15][16]. Additionally, support activities are performed in the backroom, such as break bulk of transportation units to end-user units and additional merchandise of products that have high demand or that in the selling area would quickly deteriorate (such as fruits and vegetables). Storing inventory in two locations has disadvantages because it requires permanent attention to real-time sales in order to prevent out-of-shelf (OOS) situations and lost sales. Furthermore, since in grocery retail store delivery frequencies are high, stock is needed to meet the demand for a short period of time, in contrast with most DCs.

The backroom storage area is divided into two major areas: storage area and social areas also referred as secondary departments in the literature. Storage areas, also called primary departments, are separated in three major departments, depending on the products that are stored: food, non-food and chilled areas. However, in a more detailed level, each department is organized in several sub-departments, with their own layout established. The sub-departments existence depends on the products' category and services performed. For this reason, different types of stores (hypermarkets, supermarkets, or convenience stores) require different sub-departments. An example is that hypermarkets have their own manufacture of bread, and for this reason need extra areas for ovens. Further, social areas, such as offices, restrooms and rooms, are intended for the employees of the store in order to provide them the necessary conditions to perform selling, maintenance and administrative activities.

In most companies the design of the backroom areas is mainly established on the perception of the architect and on similar stores, when it should be carefully studied based on in-store logistics, expected orders' volume of the regular activity as well as seasonal and promotional activity. To prevent these situations, designers should also rely on formal means to assist the design process, rather than follow ad hoc procedures [17].

Most performance problems associated to the design of backroom storage found in the literature are related to the lack of inappropriate architecture and store design. Despite having similar functions to a DC, backroom storage facilities have particularities that deserve a distinct analysis and can be divided in design and operational particularities. One of the main design differences from DCs is the low and irregular shape caused by the construction space as well as selling area restrictions. These warehouses are integrated in stores which are usually located in residential areas that are more expensive. Additionally, this space coexists with the selling area, which competes for the same space, and shares with it its resources (equipment, personnel, etc.). Another difference between DCs and backrooms is the low level of automation of the latter, which relies in more manual operations. Logistics processes in retail stores represent 40% of the total working hours and 40% of total retail costs due to manual activities and to the limited possibilities for using technology [15]. Another important difference is that, although the layout is divided in different areas depending on the category of the products, they all serve the same client, which is the sales area. Thus, it operates a unit serving one market with the important particularity that orders are not known and specified in advance, as occurs in DCs.

Regarding operational particularities, a big distinction between DCs and backrooms is that the latter are not as organized as the first. The differentiation factor is that personnel in charge of the backroom are the same responsible for replenishment, checkouts and help clients. This causes disorder in the backroom since it is not considered a priority as most of the efforts are attributed to the availability of products in the sales area and assisting clients in purchasing. For these reasons backroom storage areas are often neglected. Finally, backrooms accommodate technical departments, not referred in DCs, with the purpose to support the store activities, as the advertisement of the store. Backroom operational particularities are also associated with in-store operations, which account for the shelf replenishment processes that depend on the store sales and products characteristics.

In this process, the personnel travel from the backroom to the sales room, with the products about to miss, or missing, in the shelves (out-of-shelf). These activities that occur at the final stage of the supply chain of a retailer are called in-store logistics. The regular in-store operations are:

- Receipt: Products are delivered in stores by the retailer's DCs or by direct suppliers in the unloading dock, according to the delivery windows and frequencies established. Deliveries are usually made on pallets.
- Inspection: After receiving the merchandise, store personnel inspects the deliveries and confers if products delivered meet the order specifications.
- Picking products: At this stage products are separated, still in the temporary area of the dock, by the backroom departments where they are afterwards taken.
- Transporting products in the sales area: After the products are separated by store corridors, they are taken to the sales area and replenished.
- Transporting excess products to the backroom storage facilities: Excess stock that is not placed on the shelves is stored in the backroom storage area.
- Replenish of products in the shelves from the backroom: During the day and according to the necessities products are moved from the backroom storage area to the shelves.
- Handling and storing of Stock Keeping Units (SKUs) on shelves: This process includes all the activities needed in order to replenish the shelves, such as break bulk of transportation units to end user units, shelf stacking, labeling and product presentation (visual merchandising).
- Disposal/Recycling: This includes either the removal of packaging material or the disposal or recycling of damaged or broken products.

4 Backroom Design

A backroom, in similarity to a DC, consists of several departments or areas. The warehouse design phase is crucial regarding costs since it is known that warehouse costs are, to a large extent, determined at the design phase [3]. The design process of DCs is usually described by a sequence of steps. Some authors group the activities within these steps into a hierarchical framework based on a top-down approach, thus identifying strategic, tactical and operational decisions that should be considered in sequence [17]. Other authors divide the warehouse design in groups of major decisions, such as: determining the overall structure; sizing and dimensioning the warehouse and its departments; determining the detailed layout within each department; selecting warehouse equipment; and selecting operational strategies [18].

Regarding the overall structure, the main departments of a DC correspond to its major functions such as receiving, storing, packing, sorting and shipping. However, backrooms do not follow this organization since packing and shipping are not equivalent to conventional warehouses. Thus, the overall structure is unique and not a critical aspect in backrooms. Further, equipment selection in backrooms is a very limited decision since the set of equipment to be chosen from is very small and does not vary substantially between different types of stores. Finally, the generality of in-store operations are already reasonably well established and vary little between stores. For these reasons, strategy selection is not a relevant decision. Another point that we would like to stress is that conventional orders do not take place in backroom because the store clients are located in the selling area, choosing their products in the self-service display area and emptying the shelves, which is the trigger for replenishment from the backroom (backroom orders). For this reason, shipping is performed differently since products are picked and arranged in the backroom to be delivered to the sales area to replenish the shelves. Additionally, regular warehouse functions, as value added services, are also generally not performed in backrooms.

Backroom design steps, as in DCs, are interrelated and should interact during the process. Also, operational efficiency and performance are strongly affected by the design decisions. Performance evaluation is important for both warehouse design and operation, and can be assessed in terms of cost, throughput, space utilization and service levels. An alternative for performance evaluation is using Data Envelopment Analysis (DEA) techniques comparing, for instance, service levels, sales and profit of stores [19]. Using DEA the performance assessment involves a comparison among similar backroom areas located in different stores. This may allow defining targets for different sections and understanding the best warehouse layout practices. Assessing performance provides feedback about how a specific design performs compared with the requirements and how it can be improved. It is also important to state that once these decisions are established and the warehouse areas (and store) are built it can be very expensive or even impossible to make changes.

Backroom sizing and backroom layout are the key stages in the backroom design. Both these stages are influenced by and influence in-store operations and physical constraints. Backroom sizing determines the space allocation among the various departments, according to the storage requirements. Backroom layout corresponds to the detailed configuration within the backroom departments, optimizing their arrangement in the backroom. In order to solve this complex problem, a diversity of methods is described in the literature for both dimensioning and layout.

For the dimensioning stage the proposed methods range from linear to nonlinear programming formulations [3][18]; multi-attribute value functions which capture the trade-offs among different criteria [15]; heuristics to find the warehouse size [20]; integrated optimization-simulation models, which evaluate the storage shortage cost, and optimization models combined with heuristic algorithms to determine the assignment of Stock Keeping Units (SKUs) to storage areas, as well as the size of each functional areas to minimize the total handling and storage costs [21][17].

Layout problems affect warehouse performances with respect to construction and maintenance costs; material handling costs, handling machinery and energy; storage capacity, which is the ability to accommodate incoming shipments; space utilization and equipment utilization. The most used objective function is the minimization of material handling costs. However, the goals in backrooms can be different from DCs, such as the minimization of OOS, maximization of sales productivity and use of labor. Further, these objectives can be applied together in multi-objective problems. To solve the layout problem several methods are used, such as analytical formulations [18], dynamic programming, non-linear and mixed integer methods [3]. Additionally, meta-heuristics, such as genetic algorithm, are generally utilized to solve this very complex problem [21]. Simulation is also used to provide a detailed performance evaluation and a clearer view of the products flow for the resulting alternatives [20]. Additionally, a different approach to this problem is through a facility layout problem that is concerned with finding the most efficient arrangement of finite number of departments with unequal area requirements within a facility [22]. Thus, this problem can be solved through several methods depending on the objectives and the constraints defined.

5 Conclusion and Future Work

This paper draws attention to the backroom storage areas and emphasizes their importance. Furthermore, our aim is to stress the particularities of backroom storage, when compared to conventional warehouses, which support a further and separate research.

After this analysis we could conclude that the current literature focusing on warehousing is a very small fraction of the overall supply chain papers and backroom design is not discussed. Another important issue that we would like to stress is the evident lack of contributions of practical cases demonstrating potential benefits and application of academic research to real problems. Thus, cross-fertilization between the groups of practitioners and researchers appears to be limited and should be encouraged to face this challenging problem.

This discussion will, hopefully, stimulate future research in this very promising area, both from a theoretical and a practical perspective. As future research opportunities we pretend to propose a framework for designing the backroom areas and the development of a decision support system to assist designers in this complex process. Moreover, we intent to explore what is the best proportion of backroom storage areas within a store.

Acknowledgments. This work is financed by the FCT – Fundação para a Ciência e a Tecnologia (Portuguese Foundation for Science and Technology) within project UID/EEA/50014/2013.

References

[1] Hackman, S.T., Frazelle, E.H., Griffin, P.M., Griffin, S.O., Vlasta, D.A.: Benchmarking warehousing and distribution operations: an input-output approach. Journal of Productivity Analysis 16(1), 79–100 (2001)

[2] Rouwenhorst, B., Reuter, B., Stockrahm, V., van Houtum, G.J., Mantel, R.J., Zijm, W.H.M.: Warehouse design and control: Framework and literature review. European Journal of Operational Research 122(3), 515–533 (2000)

[3] Baker, P., Canessa, M.: Warehouse design: A structured approach. European Journal of Operational Research 193(2), 425–436 (2009)

[4] Hübner, A.H., Kuhn, H., Sternbeck, M.G.: Demand and supply chain planning in grocery retail: an operations planning framework. International Journal of Retail & Distribution Management 41(7), 512–530 (2013)

[5] Fernie, J., Sparks, L., McKinnon, A.C.: Retail logistics in the UK: past, present and future. International Journal of Retail & Distribution Management 38(11/12), 894–914 (2010)

[6] Trautrims, A., Grant, D.B., Fernie, J., Harrison, T.: Optimizing on-shelf availability for customer service and profit. Journal of Business Logistics 30(2), 231–247 (2009)

[7] Aastrup, J., Kotzab, H.: Forty years of Out-of-Stock research and shelves are still empty. The International Review of Retail, Distribution and Consumer Research 20(1), 147–164 (2010)

[8] Mckinnon, A.C., Mendes, D., Nababteh, M.: In-store logistics: an analysis of on-shelf availability and stockout responses for three product groups. International Journal of Logistics Research and Applications 10(3), 251–268 (2007)

[9] Kotzab, H., Teller, C.: Development and empirical test of a grocery retail instore logistics model. British Food Journal 107(8), 594–605 (2005)

[10] Berman, O., Larson, R.C.: A queueing control model for retail services having back room operations and cross-trained workers. Computers & Operations Research 31(2), 201–222 (2004)

[11] Van Zelst, S., van Donselaar, K., Van Woensel, T., Broekmeulen, R., Fransoo, J.: Logistics drivers for shelf stacking in grocery retail stores: Potential for efficiency improvement. International Journal of Production Economics 121(2), 620–632 (2009)

[12] de Koster, R., Le-Duc, T., Jan Roodbergen, K.: Design and control of warehouse order picking: A literature review. European Journal of Operational Research 182(2), 481–501 (2007)

[13] Li, Y., Cheang, B., Lim, A.: Grocery perishables management. Production and Operations Management 21(3), 504–517 (2012)

[14] Aastrup, J., Kotzab, H.: Analyzing out-of-stock in independent grocery stores: an empirical study. International Journal of Retail & Distribution Management 37(9), 765–789 (2009)

[15] Reiner, G., Teller, C., Kotzab, H.: Analyzing the Efficient Execution of In-Store Logistics Processes in Grocery Retailing-The Case of Dairy Products. Production and Operations Management 22(4), 924–939 (2013)

[16] Eroglu, C., Williams, B.D., Waller, M.A.: The Backroom Effect in Retail Operations. Production and Operations Management 22(4), 915–923 (2013)

[17] Hassan, M.M.D.: A framework for the design of warehouse layout. Facilities 20(13/14), 432–440 (2002)

[18] Gu, J., Goetschalckx, M., McGinnis, L.F.: Research on warehouse design and performance evaluation: A comprehensive review. European Journal of Operational Research 203(3), 539–549 (2010)

[19] Vaz, C.B., Camanho, A.S., Guimarães, R.C.: The assessment of retailing efficiency using network data envelopment analysis. Annals of Operations Research 173(1), 5–24 (2010)

[20] Heragu, S.S., Du, L., Mantel, R.J., Schuur, P.C.: Mathematical model for warehouse design and product allocation. International Journal of Production Research 43(2), 327–338 (2005)

[21] Gu, J., Goetschalckx, M., McGinnis, L.F.: Research on warehouse operation: A comprehensive review. European Journal of Operational Research 177(1), 1–21 (2007)

[22] Singh, S.P., Sharma, R.R.K.: A review of different approaches to the facility layout problems. The International Journal of Advanced Manufacturing Technology 30(5-6), 425–433 (2006)

The Suppliers Selection Problem: A Case Study

Mariana Costa, Cristina Requejo, and Filipe Rodrigues

Departamento de Matemática da Universidade de Aveiro, Aveiro, Portugal

Abstract. The effective supplier evaluation and purchasing processes are of vital importance to business organizations, making the suppliers selection problem a fundamental key issue to their success. We consider a complex supplier selection problem with multiple products where minimum package quantities, minimum order values related to delivery costs, and discounted pricing schemes are taken into account. Our main contribution is to present a mixed integer linear programming (MILP) model for this supplier selection problem. The model is used to solve several examples including three real case studies from an electronic equipment assembly company.

1 Introduction

The Suppliers Selection (SS) problem in Supply Chain Management represents one of the most critical tasks to be performed by the purchasing department of a business organization. The effective supplier evaluation and purchasing processes are of vital importance as the business organizations are becoming increasingly dependent on their suppliers. With the market globalization, the number of potential suppliers and the number of factors to consider when selecting suppliers increases. Therefore, an effective and efficient supplier selection process becomes very important to the success of any manufacturing or service organization [19].

We consider the supplier selection problem arising at an electronic equipment assembly company with multiple products, which is much more complex than the single product problem. For this reason, the majority of the published studies about this issue are very recent. Additionally we are considering minimum package quantities, minimum order values related to delivery costs and discounted pricing schemes. However, since we are dealing with electronic products stocking policies are not desirable.

A common approach to the SS problem with multiple products is a multi-objective approach. However, in this paper, we model the problem as a single objective optimization problem assuming price as the most important objective to the company. In [1, 5, 14, 17, 23, 24] multi-objective (mixed integer) programming approaches are proposed and frequently integrated with other approaches. Such approaches include the Analytic Hierarchy Process (AHP) ([21]), genetic algorithms ([4, 22]), the use of fuzzy concepts ([9, 12, 13, 16, 23]) including the Fuzzy TOPSIS method ([10, 11]). Together with these approaches several authors consider stochastic demands ([10, 22,

© Springer International Publishing Switzerland 2015

A.P.F.D. Barbosa Póvoa and J.L. de Miranda (eds.), *Operations Research and Big Data*,
Studies in Big Data 15, DOI: 10.1007/978-3-319-24154-8_21

24]). The most commonly used criteria are the products prices, the delivery date and the quality of products and services. Furthermore, suppliers capacity constraints, quantity discounts (frequently with cost level conditions), delivery costs and budget constraints are conditions generally considered. In [6, 14] lot-sizing and stock constraints are considered.

Our main contribution is to present a mixed integer linear programming (MILP) model to this complex SS problem with multiple products where minimum order values (MOV) related with delivery costs, minimum package quantities (MPQ) and discounted pricing schemes are taken into account. With the proposed model we aim to obtain the best supply condition for all the required products minimizing the total purchasing cost.

It is assumed that there is at least one supplier for each product and not all suppliers can supply all products.

Delivery costs are payable only in certain cases when the buyer order value does not reach certain minimum order value (MOV). In this case, if the total purchasing cost of the order to a supplier is less than its MOV, then the delivery cost, which is assumed to be fixed and not depending from the order value, has to be paid by the buyer.

The motivation for using discounted pricing schemes stems from the fact that it tends to encourage buyers to search for larger quantities and to reap operating advantages (such as economies of scale) for the buyer. From a coordination perspective, it has been shown that both the buyer and the supplier can realize higher overall profits if discounting schemes are used to set prices [18]. In [2] are listed several studies related with the SS problem when considering quantity discounts. Recently, other works were published [3, 5, 8, 15]. For each supplier and each product a set of product quantity levels is defined and each level is associated with a cost and offering some discount facility.

The remainder of the paper is organized as follows. In Section 2 we present a MILP model to the problem. In Section 3, two numerical examples and three real case studies are presented and the obtained computational results are discussed. Conclusions are drawn in Section 4.

2   Mixed Integer Linear Programming Model

In this section we present a MILP model to the SS problem with multiple products, minimum order values, discounted pricing schemes and minimum package quantities. The model presented here is similar to the model proposed in [17] however, here, conditions on the batches dimension (MPQ) are imposed and MOV related to delivery costs are considered. Furthermore, the model described herein has a single objective function minimizing the total purchasing cost.

Consider the set $S = \{1, ..., m\}$ of m suppliers and the set $P = \{1, ..., r\}$ of r products. For each supplier $s \in S$ define the set P_s of available products and for each product $p \in P$ consider the set S_p of available suppliers. To define discounted pricing schemes, define, for all $s \in S$ and all $p \in P_s$, the set $N_{sp} = \{1, ..., \lambda_{sp}\}$ of cost level conditions. To each cost level condition in set N_{sp} corresponds a product quantity level, a product unit cost and a discount facility offered by supplier s to product p when its ordered quantity ranges from the product quantity level up to the product quantity level of the next cost level condition (infinity, if last cost level condition). The λ_{sp} represents the last cost level condition and is the total number of cost level conditions that supplier s offers to product p. Therefore the total number of supplier conditions is $SC = \sum_{s \in S, p \in P_s} \lambda_{sp}$ and this number corresponds to the overall total number of possible price choices.

Consider the following parameters:

c_{spn}: unit cost of product p from supplier s at cost level condition n, for all $s \in S$, $p \in P_s$, $n \in N_{sp}$;

d_s: delivery cost associated with supplier s, for all $s \in S$;

mov_s: minimum order value (MOV) for supplier s, for all $s \in S$;

q_{spn}: product quantity level corresponds to the minimum quantity order of product p to supplier s in cost level condition n at which price c_{spn} can be considered, for all $s \in S$, $p \in P_s$, $n \in N_{sp}$. Notice that $q_{sp1} = 0$;

$pack_{sp}$: minimum package quantity (MPQ) of product p for supplier s, for all $s \in S$, $p \in P_s$;

Q_p: demand quantity of product p, for all $p \in P$.

To obtain a formulation consider the following variables:

x_{spn}: integer variables indicating the number of packages of product p ordered to supplier s in cost level condition n, $s \in S$, $p \in P_s$, $n \in N_{sp}$;

w_{spn}: integer variables indicating the number of units of product p ordered to supplier s in cost level condition n, $s \in S$, $p \in P_s$, $n \in N_{sp}$;

y_{spn}: binary variables with value 1 if cost level condition n associated with product p and supplier s is used, $s \in S$, $p \in P_s$, $n \in N_{sp}$, value 0 otherwise;

z_s: binary variables with value 1 if supplier s is used, and value 0 otherwise, $s \in S$;

t_s: binary variable with value 1 if the delivery cost associated to supplier s is supported by the buyer, 0 otherwise, $s \in S$.

To minimize the overall cost, including the total products cost and any delivery costs associated with the suppliers, the objective function of the SS model is as follows

$$Z = \sum_{s \in S, \, p \in P_s, \, n \in N_{sp}} w_{spn} c_{spn} + \sum_{s \in S} t_s d_s .$$

To satisfy all the products demand, the following constraint is considered

$$\sum_{s \in S_p, \, n \in N_{sp}} w_{spn} \geq Q_p, \qquad p \in P . \tag{1}$$

To guarantee that each product is supplied by only one supplier using only one cost level condition we have

$$\sum_{s \in S_p, \, n \in N_{sp}} y_{spn} = 1, \qquad p \in P . \tag{2}$$

The following constraint guarantees that in the case a product is supplied by a supplier using some cost level condition the quantity order should be superior to the minimum order quantity of that product necessary to qualify it for the conditions established by the cost level condition.

$$w_{spn} \geq q_{spn} y_{spn} \qquad s \in S, \, p \in P_s, \, n \in N_{sp} . \tag{3}$$

Most of the products are provided in batches, therefore the quantity order of each product should be multiple of the batch size. Thus we have

$$w_{spn} = x_{spn} pack_{sp} \qquad s \in S, \, p \in P_s, \, n \in N_{sp} . \tag{4}$$

The following constraints relate the delivery costs with the MOV. The value $\sum_{p \in P_s, \, n \in N_{sp}} \left(w_{spn} \times c_{spn} \right)$ is the total order value for supplier $s \in S$. When for a selected supplier $s \in S$, we have $z_s = 1$ the following holds. When delivery costs are to be paid by the buyer, $t_s = 1$, constraints (5) are redundant, therefore no constraints exists for the corresponding order value. Otherwise, when there are no delivery costs, $t_s = 0$, its order value (if it exists) must be greater or equal than its MOV:

$$\sum_{p \in P_s, \, n \in N_{sp}} \left(w_{spn} \times c_{spn} \right) + t_s \times mov_s \geq z_s \times mov_s \qquad s \in S \tag{5}$$

We must guarantee that when the ordered number of packages of a product to a supplier for a certain cost level condition is zero, then the corresponding cost level condition is not used.

$$y_{spn} \leq x_{spn} \qquad s \in S, \, p \in P_s, \, n \in N_{sp} \tag{6}$$

To guarantee that the corresponding cost level condition is used, when the ordered number of packages is non null, we have

$$My_{spn} \geq x_{spn} \qquad s \in S, p \in P_s, n \in N_{sp} \qquad (7)$$

with M being the maximum number of packages that can be ordered. We can use
$$M = \left\lceil \max\left\{ Q_p, \frac{mov_s}{c_{spn}}, q_{spn} \right\} / pack_{sp} \right\rceil, \quad \forall s \in S, p \in P_s, n \in N_{sp}.$$

Next we establish that if a cost level condition is used for a supplier, then the corresponding supplier must also be used:

$$\sum_{p \in P_s, n \in N_{sp}} y_{spn} \leq SC \times z_s \qquad s \in S. \qquad (8)$$

If no cost level condition is used for a supplier, then this supplier is not the solution:

$$\sum_{p \in P_s, n \in N_{sp}} y_{spn} \geq z_s \qquad s \in S \qquad (9)$$

Finally we have the constraints on the variables values. For the integer variables

$$x_{spn}, w_{spn} \; integer \qquad s \in S, p \in P_s, n \in N_{sp} \qquad (10)$$

and for the binary variables

$$z_s, t_s, y_{spn} \in \{0, 1\} \qquad s \in S, p \in P_s, n \in N_{sp}. \qquad (11)$$

The MILP model for the SS problem is as follows: min Z subject to (1)-(11).

The number of constraints and variables in the model can be expressed according to the number SC and the cardinality of sets S and P. The model uses $2|P| + 3|S| + 4SC$ constraints and $2|S| + 3SC$ variables.

3   Numerical Examples and Case Studies

In this section we describe and analyze two small examples and three real case studies from an electronic equipment assembly company. With the small examples we explore and describe characteristics of the problem. The case studies help to validate the model tailored for the assembly company. We use the software Xpress-Ive 64 bits, version 7.6., to solve the examples by using the model and we run the software on a personal computer with Intel(R) Core(TM) i5-2410M CPU @ 2.30 Ghz and 6 GB of RAM.

The first example has four suppliers, four products and eighteen cost level conditions. The characteristics of this example are displayed in the Table A1 presented in Appendix. The required quantities for the products are 100, 135, 129 and 115 units, respectively and the obtained solution is displayed in Table 1.

Table 1 Summary of the results of the first example

Supplier	Product	Quantity	Order Value	Delivery cost
1	-	-	-	-
2	4	120	108	no
3	1,3	100,150	100	no
4	2	150	45	yes

The ordered quantities of the products 1, 2, 3 and 4 are, respectively, of 100, 150, 150 and 120 units which are higher than the required quantities for products 2, 3, and 4 to satisfy the MPQ quantities. Supplier 1 receives no order, therefore is not used. Products 1 and 3 are provided by supplier 3 and the total order value for this supplier is greater than its MOV. For this reason, there is no delivery cost associated to this supplier. The same happens for supplier 2, where product 4 is ordered. The reverse situation occurs for supplier 4, as the total order value is less than its MOV thus the buyer has to pay delivery costs. Notice that 120 units of product 4 are ordered at cost 0.9. The order quantity of this product allows a unit cost of 0.8 corresponding to the next cost level condition, however, in this case, the MOV would not be achieve, the buyer would have to pay the delivery cost and the total cost would be higher. On the other hand, the ordered quantity of product 2 to supplier 4 is lower than its MOV, however there is advantage on paying the delivery cost instead of reaching its MOV. This would imply the order of 250 units of product 2 at a higher cost. The total cost for this example is of 268.

The second example has three suppliers and five products. The table with the specifications of this example is Table A2 in Appendix. The required quantities for the products are 988, 480, 670, 200 and 775, respectively. The solution is displayed in Table 2.

Table 2 Summary of the results of second example

Supplier	Product	Quantity	Order Value	Delivery cost
1	2, 3, 4	520, 720, 200	50	no
2	-	-	-	-
3	1,5	1000, 805	80.25	no

The ordered quantities of the products 1, 2, 3, 4 and 5 are, respectively, of 1000, 520, 720, 200 and 805 units which are higher than the required quantities for products 1, 2, 3, and 5. There are different reasons for this fact. Product 5 has to satisfy the MPQ quantity. For products 1, 2 and 3 the reason is to satisfy MOV. Supplier 2 is not used. The products 2, 3 and 4 are provided by supplier 1 and the total order value for this supplier is equal to its MOV. For this reason, there is no delivery cost associated to this supplier. The same happens with supplier 3, where products 1 and 5 are ordered. The total cost is 130.25.

Next we present three real case studies, named projects 1, 2 and 3, of the SS problem from an electronic equipment assembly company. In this company, once for each project, the SS process is manually performed by workers. The supply conditions for which the unit price is the smallest possible are chosen, whereby the process time is slow. For the three case studies the number of available suppliers, the number of required products and the number of available supply conditions (related with quantities discounts) are displayed in Table 3.

Table 3 Number of suppliers, products and supply conditions for the case studies

	Suppliers	Products	Supply Conditions
Project 1	48	106	616
Project 2	40	35	275
Project 3	49	58	445

In Table 4 we display the project cost obtained manually by the company in column "**Cost reference**". The remaining columns refer to the application of the MILP model to solve the problem. The obtained total cost of the project is displayed in column "**Model cost**", the computational time used in column "**Execution Time**" and the number of constraints and variables are displayed in columns "**Constraints**" and "**Variables**", respectively.

Table 4 Results for three real case studies

	Cost Reference	Model Cost	Execution time	Constraints	Variables
Project 1	39309	39299	0.025	2820	1944
Project 2	24319,2	24206,7	0.008	1290	905
Project 3	27044,2	26623,8	0.012	2043	1433

From Table 4 we conclude that the results obtained by using the proposal model are better than the results obtained manually by the company. Moreover, we may conclude that the execution time of the model increases slightly with the number of available supply conditions, although, these times are much smaller than the times used by the workers of the company to obtain the solution manually.

4 Conclusions

We address the supplier selection problem for several products considering minimum order values (MOV) related with delivery costs, minimum package quantities (MPQ) and discounted pricing schemes. A mixed integer linear programming model is proposed that minimizes the total cost, the criteria elected as the most important. We present two small examples to describe and analyze some characteristics of the problem and three case studies.

Acknowledgements. The research of the author Cristina Requejo was partially supported by Portuguese funds through the *Center for Research and Development in Mathematics and Applications* (CIDMA) and FCT, the Portuguese Foundation for Science and Technology, within project UID/MAT/04106/2013.

References

[1] Arikan, F.: Multiple Objective Fuzzy Sourcing Problem with Multiple Items in Discount Environments. Mathematical Problems in Engineering, 1–14 (2015)

[2] Benton, W.C., Park, S.: A classification of literature on determining the lot size under quantity discounts. European Journal of Operational Research 92, 219–238 (1996)

[3] Burke, G., Carrillo, J., Vakharia, A.: Heuristics for sourcing from multiple suppliers with alternative quantity discount. European Journal of Operational Research 186, 317–329 (2006)

[4] Che, Z.H., Wang, H.S.: Supplier selection and supply quantity allocation of common and non-common parts with multiple criteria under multiple products. Computers & Industrial Engineering 55, 110–133 (2008)

[5] Dahel, N.E.: Vendor selection and order quantity allocation in volume discount environments. Supply Chain Management 8, 335–342 (2003)

[6] Esmaeili, D., Kaazemi, A., Pourghannad, B.: A two-level GA to solve an integrated multi-item supplier selection model. Applied Mathematics and Computation 219, 7600–7615 (2013)

[7] Feng, B., Fan, Z.P., Li, Y.: A decision method for supplier selection in multiservice outsourcing. International Journal of Production Economics 132, 240–250 (2011)

[8] Ghodsypour, S.H., O'Brien, C.: The total cost of logistics in supplier selection, under conditions of multiple sourcing multiple criteria and capacity constraint. International Journal of Production Economics 73, 15–27 (2001)

[9] Jolai, F., Yazdian, S., Shahanaghi, K., Khojasteh, M.: Integrating fuzzy TOPSIS and multi-period goal programming for purchasing multiple products from multiple suppliers. Journal of Purchasing and Supply Management 17, 42–53 (2011)

[10] Kara, S.: Supplier selection with an integrated methodology in unknown environment. Expert Systems with Applications 38, 2133–2139 (2011)

[11] Kilic, S.: An integrated approach for supplier selection in multi-item/ multi-supplier environment. Applied Mathematical Modelling 37, 7752–7763 (2013)

[12] Nazari-Shirkouhi, S., Shakouri, H., Javadi, B., Keramati, A.: Supplier selection and order allocation problem using a two-phase fuzzy multi-objective linear programming. Applied Mathematical Modelling 37, 9308–9323 (2013)

[13] Razmi, J., Maghool, E.: Multi-item supplier selection and lot-sizing planning under multiple price discounts using augmented constraint and Tchebycheff method. International Journal of Advanced Manufacturing Technology 49, 379–392 (2010)

[14] Rezaei, J., Davoodi, M.: Multi-objective models for lot-sizing with supplier selection. International Journal of Production Economics 130, 77–86 (2011)

[15] Rubin, P.A., Benton, W.C.: Evaluating Jointly Constrained Order Quantity Complexities for Incremental Discounts. European Journal of Operational Research 149, 557–570 (2003)

[16] Torabi, S., Hassini, E.: Multi-site production planning integrating procurement and distribution plans in multiechelon supply chains: an interactive fuzzy goal programming approach. International Journal of Production Research 47, 5475–5499 (2009)

[17] Wadhwa, V., Ravindran, A.R.: Vendor selection in outsourcing. Computers & Operations Research 34, 3725–3737 (2007)

[18] Wang, Q.: Discount Pricing Policies and the Coordination of Decentralized Distribution Systems. Decision Sciences 36, 627–646 (2005)
[19] Ware, N., Singh, S., Banwet, D.: Supplier selection problem: A state-of-the-art review. Management Science Letters 2, 1465–1490 (2012)
[20] Weber, C.A., Current, J.R., Desai, A.: An optimization approach to determining the number of vendors to employ. Supply Chain Management 5, 90–98 (2005)
[21] Xia, W., Wu, Z.: Supplier selection with multiple criteria in volume discount environments. Omega 35, 494–504 (2007)
[22] Yang, P.C., Wee, H.M., Pai, S., Tseng, Y.F.: Solving a stochastic demand multiproduct supplier selection model with service level and budget constraints using Genetic Algorithm. Expert Systems with Applications 38, 14773–14777 (2011)
[23] Zarandi, M.H.F., Saghiri, S.: A comprehensive fuzzy multi-objective model for supplier selection process. In: The 12th IEEE International Conference on Fuzzy System, FUZZ 2003, vol. 1, pp. 368–373 (2003)
[24] Zhang, G., Ma, L.: Optimal acquisition policy with quantity discounts and uncertain demands. International Journal of Production Research 47, 2409–2425 (2009)

Appendix

Table A1. Delivery costs and MOV associated with suppliers, list of product for each supplier, quantities discounts and MPQ for these products, relatively example 1.

Supplier S	d_s	MOV_s	Product p	$pack_{sp}$	q_{spn}	c_{spn}
1	10	50	1	20	0	0.2
					200	0.15
					400	0.1
			2	40	0	0.5
					200	0.4
			4	100	0	0.8
2	20	100	2	1	0	0.6
			4	20	0	0.9
					100	0.8
					300	0.7
3	10	60	3	30	0	0.5
			1	50	0	0.25
					250	0.175
					500	0.1
4	15	65	3	25	0	0.6
					250	0.5
					500	0.4
			2	50	0	0.3

Table A2. Delivery costs and MOV, MPQ associated with suppliers, list of product for each supplier and quantities discounts for this products, relatively example 2.

Supplier S	d_s	MOV_s	Product p	$pack_{sp}$	q_{spn}	c_{spn}
1	10	50	1	20	0	0.1
					1000	0.08
			2	40	0	0.01
					500	0.007
			3	30	0	0.04
			4	100	0	0.08
			5	1	0	0.09
					500	0.07
2	20	80	2	20	0	0.015
					400	0.01
			3	10	0	0.05
					500	0.04
			4	50	0	0.085
					300	0.06
					700	0.04
3	15	60	1	25	0	0.5
					750	0.04
			2	20	0	0.02
					1000	0.005
			5	35	0	0.095
					300	0.05

An Integrated Decision Support System to Assess the Sustainability of Demand Responsive Transport Systems

Ana Cecília Ribeiro, Maria Sameiro Carvalho, and José Telhada

Centro Algoritmi, Escola de Engenharia, Universidade do Minho,
Campus de Gualtar, 4710-057 Braga, Portugal
{ana.dias,maria.carvalho,jose.telhada}@algoritmi.uminho.pt

Abstract. The provision of conventional public passenger transport services in rural areas have shown to be very inefficient and ineffective due to low levels of population density and high spatial and temporal dispersion. Demand Responsive Transport (DRT) systems have been seen as an interesting alternative solution, adopted in several countries as a way to increase user's mobility, providing flexible transport services to meet trip requests and mitigate social exclusion. However, some DRT systems have revealed to be inadequate or even unsustainable, mainly because they are highly dependent on the correct tuning of organizational and functional parameters, namely the level of flexibility of schedules, routes and stop locations. Despite the existence of a vast literature concerning DRT systems, very few contributions have been put forward concerning comprehensive approaches to tackle these problems and to assess sustainability. This paper proposes a new integrated multi-disciplinary decision support system (DSS) to help decision-makers design and plan DRT systems, and assess their sustainability. The proposed DSS comprises analytical and simulation tools and consists of a procedure that iteratively simulates and evaluates alternative specifications of the system until an adequate and sustainable solution is encountered. The new tool has been applied to a case study in a rural area of the northwest of Portugal, and preliminary results suggest that the proposed approach was able to identify a set of conditions under which a DRT system can be sustainable.

Keywords: Demand Responsive Transport, Decision Support System, Simulation.

1 Introduction

Conventional public passenger transport services in low density rural areas tend to be highly inefficient for operators because of low occupancy rates of their vehicles. Simultaneously, they are not effective because do not fit well demand due to rigidity and scarcity of their services in terms of routes and schedules. Demand Responsive Transport (DRT) systems have been seen as an interesting alternative

© Springer International Publishing Switzerland 2015   185
A.P.F.D. Barbosa Póvoa and J.L. de Miranda (eds.), *Operations Research and Big Data,*
Studies in Big Data 15, DOI: 10.1007/978-3-319-24154-8_22

solution, adopted in several countries as a way to increase user's mobility, providing flexible transport services to meet trip requests and mitigate social exclusion. According to [1], the potential advantages of such systems include: *"potential to increase public transport patronage; integration between current fixed route and flexible transport systems to achieve more 'holistic' transport solutions; ability to serve areas with demand too high for door-to-door, but too low for fixed route services; making public transport more attractive to choice users, hence reducing car use and associated problems."*

There is however some issues concerning DRT scheme design and evaluation that require further developments. In fact, the lack of adequate tools to support some of the strategic and tactic level decisions that must be made at the design phase have already been identified. In the last decades a few number of contributions have been developed to overcome some of these problems [2-3]. In [4], a simulation tool has been proposed to analyze the impact of time windows on fleet size, and centralized versus decentralized operative scenarios. The lack of guidance on the most appropriate service and system design was pointed out by [2]. The role of decisions support systems to support different decisions that need to be taken in the design of complex systems have also been addressed [5]. However, as far to our knowledge is concerned, there are no contributions in the literature dealing with strategic, tactical and operational design issues of DRT systems, in an integrated way.

In this research, we propose a Decision Support System (DSS) that will allow achieving better planning decisions evaluating alternative scheme designs and management strategies of DRT systems, and assess their sustainability, prior to their implementation. The DSS is based on a procedure that iteratively simulates and evaluates alternative solution designs until an adequate and sustainable solution is encountered. The simulation component consists on an event-driven day-to-day and within-day dynamic model, which emulates real-world customers' behavior and vehicles movements.

The remainder part of the paper is organized as follows. Section 2 presents a literature review on DRT systems, highlighting main conceptual and design issues. Section 3 presents the new DSS model and a discussion of the main issues it addresses. Section 4 illustrates the applicability of the DSS by reporting and discussing the main results of a case study. Finally, Section 5 reports the main conclusions and final considerations regarding future developments.

2   Literature Review

Flexible Transportation Systems (FTS) include a widespread range of non-conventional passenger or freight transportation services. The flexibility of these transportation services refers to the following features: 1) space (the route and stop locations), 2) time (the schedule), 3) types of vehicles used, 4) the booking system, and 5) ways of payment, or combinations of these features. A DRT system is a special type of FTS devised only to provide transport on demand from users (passengers), as a mean to mitigate mobility problems of dispersed residents and thus combat their functional social exclusion [2, 6].

In general, trip requests are made by telephone dialing to a travel dispatch center (TDC), during a pre-defined time window; there are also cases where users can also use a messaging system (SMS) or a web portal. The TDC coordinates a fleet of vehicles with communication technologies. Its implementation typically involves the use of information and communication technologies comprised in an operational DSS [2, 7].

A heterogeneous fleet of vehicles are assigned to trip requests by the DSS that incorporates some rationality to the system throughout the use of vehicle routing algorithms, obtaining adequate transport solutions according to the area characteristics and demand patterns. The use of a DSS with these and others rational features (typically classified as an Intelligent Transportation System, ITS), has endorsing collaboration of multiple service providers, and it has enhancing flexibility and popularity by providing intelligent solutions to process trip requests, improving cost-effective performance of DRT services [2, 6-9].

According to [5], a DSS which combines the use of models, analytical techniques, data access and retrieval functions, by using advanced differentiated technologies, will have the advantage (in relation to traditional DSS) of providing consistent, coordinated, active and global support for the various managers/analysts on their different levels of decision-making processes.

Most of the work developed so far identifies as key success factors for the sustainability of a DRT a set of strategic level decisions concerned with system conceptualization issues: specific policy goals, target market, area factors (e.g. population density, income level, demographic details, land use pattern), stakeholders involved, the regulatory and financial model and fare structure. More specifically, several authors [e.g. 4, 6, 10-11] had identified most critical decisions to be taken: level of flexibility to adopt (routes, timetables, time-windows), operating rules, resources (fleet, drivers, informatics, TDC center and staff), fare definition, integration level and system evaluation. One of the main issues is concerned with DRT cost per passenger, which is a relatively high cost when compared to taxis, and thus fare revenue is unlikely to cover costs. Therefore, these projects tend to close after the pilot stage or when funding schemes end [3]. Additionally, it is fundamental to estimate the impact that some of these decisions have in terms of overall system performance (e.g. cost, quality) and several indicators can be used to evaluate a DRT service [1, 4, 12-13].

Despite the high number of real cases reported in the literature concerning the implementation of DRT schemes, only few authors report the methodology that had been adopted in the planning and implementation process. A sequential framework have been proposed in the context of some EU funded projects [e.g. 16], which consists on a two steps approach: design and planning phase and the implementation phase. In this context, Silva [15] proposed that at the initial stage, the existing transportation system must be evaluated, mobility problems must be clearly identified, and the adequacy of a flexible transportation system should be preliminary assessed as well as the existing regulatory and legal framework.

The following stage, according to the same approach [15], consists of designing the new DRT system, addressing a well-defined set of issues: how to provide a sustainable passenger transport service, the definition of its objectives in the context of a wider system general policies, organizational model, the definition of a financing plan, and its general functioning rules.

3   The New Integrated Decision Support System

This section describes a DSS that can be used in the context of the general framework presented by [15] for designing a sustainable DRT system. Sustainable transportation systems are an urgent issue and new approaches are required to incorporate social, environmental and economic concerns in the design of new transportation schemes. The proposed DSS is based on a simulation approach and it is intended to provide support to questions concerning tactical and operating rules such as flexibility level (what level(s) of flexibility should be adopted, in terms of stops, timetables, frequencies, time-windows?), resources level (what resources are going to be used (fleet, drivers)?); service level (what rules should be adopted?), by delivering a set of measures that allow the assessment of system sustainability in terms of economic, social and environmental aspects. The new tool (DSS) complements the general designing and planning framework proposed by [15], by providing a detailed technical specification (and computational implementation) of the analytical and simulation models and sub-models to quantitatively address the questions formulated in above.

The DSS (Figure 1) is based on a micro-simulation model to reproduce the functioning of the DRT under alternative policies, and estimate the corresponding key performance indicators (KPI), helping decision-makers to assess their impact in terms of future viability and sustainability. A set of routines with optimization, simulation and statistical methods will perform the necessary operations to generate rational routes and schedules required to satisfy the within-day dynamics of trip requests on a daily basis, and for a pre-specified (high) number of days ("simulation runs") for a given level of resources (e.g. number and capacity of vehicles), a given set of operating rules (e.g. route and time flexibilities), and a given operating scenario (e.g. level of adhesion of population). The operational results of the simulation are stored in a database. Then, the assessment or analyst model extracts and converts such results to adequate and meaningful KPIs to aid the decision maker to assess the simulated solution (the set of inputted resources and operating rules). Since the outcome of the performance evaluation is highly dependent on the DRT specification (in terms of operational parameters), so this framework must involve an iterative approach between the simulator and the assessment module until a final suitable and sustainable solution is found. In each iteration, the decision maker should readjust the solution to be simulated next.

The simulation component of the DSS comprises two main models: a demand-side model, implemented as a travel request generator and users' behavior simulator, and a supply-side simulator that simulates the functioning DRT services, including the system operating parameters and the resources (vehicle operations). The main demand-side events are trip reservation, trip cancelation, user arrival to

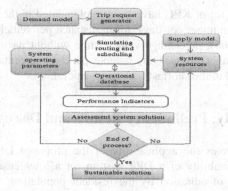

Fig. 1 General architecture of the DSS

stops (origin and destination), user no-show and non-reserved shows at stops. The main supply-side events are trip planning events (e.g. hourly and daily vehicle scheduled planning of previous reservations, real-time acceptances or rejections and re-planning), vehicle departure and arrival to stations, stops and strategic waiting and interface points. Vehicle scheduling is performed by a vehicle routing problem (VRP) module based on a set of efficient heuristics that were adapted from [16-17] for solving the variants of the problem that can arise in practice (e.g. single origin-multiple destination, dial-a-ride).

Travel requests are generated using statistical models, based on socio-economic characteristics of the resident population (from Census), from domiciliary questionnaires and local authorities interviews, as well as acquired knowledge about the main attracting zones (workplaces, schools, hospitals, markets, general services, transfer stops to inter-urban transport services, etc.).

Both demand and supply models are based on a micro-simulation event-driven approach. Some degree of freedom (in choosing any random destination) is also modeled by a given probability (as a model parameter) and performed by using a gravity model. User time arrivals at departure stops and vehicle movements are simulated by other independent random generators (and probabilistic laws). Average travel conditions, such as travel times, are taken from origin-destination (OD) trip times previously stored in the system database and obtained by invoking Google Maps internet services (shortest route between two points).

The assessment module will provide the decision maker with a large set of KPIs to assess solution viability and sustainability, encompassing the three important dimensions for decisions in transport investments, that is social, economic and environmental dimensions. The module is powered by the Business Intelligence type framework proposed by [18] that comprises different methodologies: simple and advanced statistical reporting and inference techniques, data mining and operational research inference and prospective models. The result of the assessment process will provide guidelines and the required feedback to adjust system resources and operating parameters for further analyses. KPIs are illustrated in the form of tables, graphs and reports. KPIs helps DRT managers (TDC coordinators, system designers and analysts) redefine and restructure system definitions and budgets.

An appropriated set of KPIs have been implemented in the DSS, such as: fleet size (mean vehicle number per day); mean distance per vehicle; mean occupation rate (total and partial); percentage vehicle time without passengers; mean travel distance (per passenger); passengers number per day; percentage increase in time and distance related to the shortest path (per vehicle).

4   Case Study: Preliminary Results and Discussion

This section illustrates the applicability of the proposed DSS in assessing the viability and sustainability of a DRT system for a low-density rural area under different scenarios of adhesion by the resident population. Since sustainability encompasses a wide set of issues and questions, and their integrated and weighed analysis, the results and conclusions herein reported should not be taken as a complete or definitive study concerning the implementation of the DRT in the territory, but merely as an illustrative test (and validation test) of the applicability of the DSS in designing such a DRT transportation system. The case study is based on a rural area in Minho region, north of Portugal (Terras de Bouro, a county with the lowest population density of the region). The DSS was applied to assess models behavior whenever testing different scheme designs, such as different demand levels, different degrees of route and timetable flexibility, among others and assess sustainability of the system.

Table 1 reports the values of KPIs for different values of adhesion defined as the percentage of the adult population of the study area. As expected, the required fleet size (number and larger vehicles) increase as demand level (and trip requests) increases. Consequently, all the KPIs increase with the growth of adhesion, except the percentage of vehicle time without passengers, which decreases, as expected. For example, deviations from the shortest paths increases as demand increases, because more different requests are likely to share the same trip (with larger vehicles) and therefore more extra kilometers are needed to collect all passengers. On average, regular public transport (PT) takes up more 82.8% of the time that would take a passenger to go from his origin to his destination by car (shortest path), which corresponds to approximately 2.59 minutes in commuting [19]. In this regard, and according to the penultimate row of Table 1, the DRT service would outperforms actual PT service for adhesions of 5% or less.

Table 1 KPIs for different levels of adhesion (% total adult population)

Adhesion	1%	2%	5%	10%
Average fleet size (number of vehicles: 4 to 8 seats)	3.4	5.2	9.4	14.5
Mean distance per vehicle (km)	120	147	175	194
Mean occupation rate (total)	10.4	12.7	15.8	17.6
Mean occupation rate (partial)	27.3	30.5	33.2	33.8
Percentage of vehicle time without passengers (%)	71	67.6	61.7	57.2
Mean passenger number (per day)	20.8	43	111	221
Mean travel distance per passenger (km)	8.6	9	9,5	9,9
% increase in waiting+travel time related to shortest path (per vehicle)	46.1	59.2	82	93.7
% increase in distance related to the shortest path (per vehicle)	16.7	20.1	27.5	30.1

Table 2 reports the outstanding cost (to pay by users) in function of the financing level and adhesion. These values were estimated by [19] whose calculations were based on the case of the DRT of Mação [20], a village and municipally in the center of Portugal, where the average fare per trip is 3.35€ with 70 to 90% financing (see [19] for details). Considering the financing of 70%, the annual cost of service varies from about 18,000€ (1% adhesion) and 104,000€ (10% adhesion), where the respective average costs per passenger decrease with increasing demand to 3.18€ to 1.79€ and may be competitive with regular PT with increased adhesion.

Table 2 Outstanding operating cost (per year and user-trip) by financing level and by adhesion

Adhesion	Financing percentage of total operating costs											
	No financing		20%		40%		60%		70%		80%	
	€/year	€/u-trip	€/year	€/u-trip	€/year	€/u-trip	€/year	€/u-trip	€/year	€/u-trip	€/year	€/u-trip
1%	58,000	10.58	46,000	8.47	35,000	6.35	23,000	4.23	18,000	**3.18**	12,000	**2.12**
2%	100,000	8.82	80,000	7.06	60,000	5.29	40,000	**3.53**	30,000	2.65	20,000	1.76
5%	203,000	7.05	162,000	5.64	122,000	4.32	81,000	**2.82**	61,000	2.11	41,000	1.41
10%	345,000	5.97	276,000	4.78	207,000	**3.58**	138,000	2.39	104,000	1.79	69,000	1.19

Despite the benefits suggested by these analyses, this study has limitations in access to the complete set of information for the calculation of all cost components for performing the financial sustainability analysis of DRT services, so further data gathering and analyses will be necessary to ascertain a more effective financial viability of the system.

Table 3 shows the estimations of changes in emissions between DRT services when compared to the existing regular PT service in the study area [19]. In all cases, it is expected that the DRT system would outperform the actual system.

Table 3 Change in emissions between the idealized DRT service and the actual PT service

Adhesion (and vehicle fleet)	CO2	CO	NOX	NMVOC	PM
1 % (4 vehicles of 4 seats)	-96.1%	-98.3%	-98.5%	-98.6%	-95.4%
2% (6 vehicles of 4 seats)	-92.7%	-96.8%	-97.2%	-97.3%	-91.5%
5% (11 vehicles of 4 to 6 seats)	-84.2%	-93.1%	-93.9%	-94.3%	-81.5%
10% (16 vehicles of 4 to 6 seats, and 1 vehicles of 8 seats)	-72.5%	-86.8%	-89.2%	-89.1%	-66.9%

5   Conclusions

To date, very little relevant research has been conducted to address main planning issues of DRT systems, namely tools to support designing decisions.

In this research critical designing issues are identified and a DSS is proposed to provide different levels of decision-support to enable managers to perform systematic analysis leading to intelligent strategic solutions in order to define a sustainable DRT system before undertaking its implementation. The DSS integrates a rich set of analytical models, simulation techniques and other advanced technologies (e.g., internet services, GIS, intelligent agents, advanced statistical, OR and artificial intelligent tools), thus enhancing its capacity and efficiency of analysis and assessment. The DSS will provide decision makers with an insight of different systems design performance allowing for a better decision making process.

Based on the demand and supply models representing the main characteristics of the transportation system in a particular area, for a given level of resources and a given set of operating rules, the DSS simulates the real system and a set of performance indicators. By evaluating alternative scenarios, it allows choosing the most adequate level of resources and the most suitable set of working rules, which constitutes one of the key elements of this framework since it will provide essential information for the system evaluation process.

Preliminary results of its application to a study case of a rural area in the northwest of Portugal suggests that the operationalization of a DRT system can be sustainable depending on, among other factors, the level of adhesion and the financial support. Additional validation tests are currently being undertaken and some more will be performed, along with parameterization tests: spots of population concentration within the different counties of the study area; DRT system integration with regular transport service; flexibility of services as a function of economic efficiency, costs effectiveness and resources availability. Other studies will include the DSS overall validation and the proposal of a DRT specification at the study area based on the application of the DSS and on the basic principles of project viability and sustainability. Moreover, the present framework needs further validation (and transferability) studies for medium and large size areas of implementation and for different typologies of territory (e.g., in terms of demand densities and dispersion).

References

[1] Ferreira, L., Charles, P., Tether, C.: Evaluating flexible transport solutions. Transportation Planning and Technology 30, 249–269 (2007)

[2] Brake, J., Nelson, J., Wright, S.: Demand responsive transport: towards the emergence of a new market segment. Journal of Transport Geography 12, 323–337 (2004)

[3] Enoch, M., Ison, S., Laws, R., Zhang, L.: Evaluation study of demand responsive transport services in Wiltshire. Transport Studies Group Department of Civil, Loughborough University, UK (2006)

[4] Quadrifoglio, L., Li, X.: A methodology to derive the critical demand density for designing and operating feeder transit services. Transportation Research B 43, 922–935 (2009)

[5] Liu, S., Duffy, A., Whitfield, R., Boyle, I.: Integration of decision support systems to improve decision support performance. Knowledge Information Systems 22, 261–286 (2010)

[6] Mulley, C., Nelson, J.: Flexible transport services: a new market opportunity for public transport. Research in Transportation Economics 25, 39–45 (2009)

[7] Ambrosino, G., Nelson, J., Romanazzo, M. (eds.): Demand responsive transport services: towards the Flexible Mobility Agency. ENEA, Rome (2004)

[8] Brake, J., Mulley, C., Nelson, J., Wright, S.: Key lessons learned from recent experience with Flexible Transport Services. Transport Policy 14, 458–466 (2007)

[9] Nelson, J., Phonphitakchai, T.: An evaluation of the user characteristics of an open access DRT service. Research in Transportation Economics 34, 54–65 (2012)

[10] Parragh, N., Doerner, K., Hartl, R.: A survey on pickup and delivery problems, Part II: Transportation between pickup and delivery locations. Journal für Betriebswirtschaft 58, 81–117 (2008)

[11] Gomes, R.: Dynamic vehicle routing for demand responsive transportation systems. PhD Thesis, Faculty of Engineering of the University of Porto, Portugal (2012)

[12] Brake, J., Nelson, J.: A case study of flexible solutions to transport demand in a deregulated environment. Journal of Transport Geography 15, 262–273 (2007)

[13] Palmer, K., Dessouky, M., Zhou, Z.: Factors influencing productivity and operating cost of demand responsive transit. Transportation Research Part A 42, 503–523 (2008)

[14] ARTS Consortium. ARTS: Actions on the integration of rural transport services (2002). http://www.ruraltransport.net/index.phtml (last accessed March 2013)

[15] Silva, J.: Conceptualization of a flexible transport system for elderly and disable. Dissertation in Industrial Engineering. University of Minho, Guimarães, Portugal (2012)

[16] Xiang, Z., Chu, C., Chen, H.: A fast heuristic for solving a large-scale static dial-a-ride problem under complex constraints. European Journal of Operational Research 174, 1117–1139 (2006)

[17] Kim, T., Haghani, A.: Model and Algorithm Considering Time-Varying Travel Times to Solve Static Multidepot Dial-a-Ride Problem. Transportation Research Record 2218, 68–77 (2011)

[18] Telhada, J., Dias, A., Sampaio, P., Pereira, G., Carvalho, M.: An integrated simulation and BI framework for designing and planning demand responsive transport systems. In: Proc. 4th Int. Conf. on Computational Logistics, Copenhangen, Denmark, September 24-26, pp. 98–112 (2013)

[19] Ribeiro, A.: Support tool for flexible system design of public passenger transport. PhD Thesis in Industrial and Systems Engineering. University of Minho, Guimarães, Portugal (2015)

[20] Mação. Projeto de transporte a pedido na área do médio Tejo (2014). http://transporteapedido.mediotejo.pt/ (last access June 20, 2014)

GPU-Based Computing for Nesting Problems: The Importance of Sequences in Static Selection Approaches

Pedro Rocha[1], Rui Rodrigues[2], A. Miguel Gomes[2], and Cláudio Alves[3]

[1] INESC TEC, Porto, Portugal
[2] INESC TEC and Faculdade de Engenharia, Universidade do Porto, , Porto, Portugal
[3] Escola de Engenharia, Universidade do Minho, Braga, Portugal

Abstract. In this paper, we address the irregular strip packing problem (or nesting problem) where irregular shapes have to be placed on strips representing a piece of material whose width is constant and length is virtually unlimited. We explore a constructive heuristic that relies on the use of graphical processing units to accelerate the computation of different geometrical operations. The heuristic relies on static selection processes, which assume that a sequence of pieces to be placed is defined *a priori*. Here, the emphasis is put on the analysis of the impact of these sequences on the global performance of the solution algorithm. Computational results on benchmark datasets are provided to support this analysis, and guide the selection of the most promising methods to generate these sequences.

1   Introduction

Given their practical and theoretical relevance, cutting and packing problems have deserved the attention of both operations research and computer science practitioners [8]. The general problem consists in finding the best way to place a set of items (pieces) on a larger object (board). The feasibility of a solution may be subject to different constraints. In the simplest form of the cutting and packing problem, both the pieces and the board are 1-dimensional objects and the only constraint that applies is related to the size of the board, which should not be exceeded. On higher dimensional problems, other constraints become more critical such as the non-overlapping of the pieces and the fact that pieces must be placed inside the board (although these constraints also arise in 1-dimensional settings, they are treated very easily in these cases). Furthermore, in real settings, many different specific operational constraints may be considered, such as the existence of conflicts between the pieces or the limitation of the rotations applied to the pieces. One interesting aspect is that they may also arise as subproblems of other integrated optimization problems, as for example in transportation problems where the capacity of the vehicles are strong constraints.

In this paper, we address the so-called *nesting problem* or irregular strip packing. The problem is a 2-dimensional packing problem involving pieces whose contour may be irregular. Here, irregularity contrasts with many other cases studied in the literature where only specific convex shapes are considered (squares, rectangles, circles). In our case, the shapes may have concavities, and they are not restricted to any particular family of polygons. The board is a rectangular strip representing a piece of material whose interior is homogeneous, has fixed width, and virtually unlimited length. Given these definitions, the optimization problem consists in finding the positions where the pieces should be placed such that they are completely inside the board, do not overlap, and the total length of the used strip is minimized. The problem is clearly NP-hard [4], and, in practice, it remains challenging even for small datasets due to its combinatorial and geometric nature. Different types of approaches have been described in the literature [2, 3, 5]. Here, we explore the potential of graphical processing units (GPU) to accelerate the computation of some geometrical operations, and we study the importance of good sequences of pieces in approaches based on greedy constructive heuristics. To the best of our knowledge, this is the first time that GPU are used to evaluate the quality of layouts from a representation of the solutions based on sequences of pieces. The objective of our study is mainly experimental. It aims to provide insights towards the development of efficient approaches from similar platforms.

The paper is organized as follows. The second section presents the basic definitions and concepts related to the nesting problem. Third section describes the solution framework adopted in our study, and discusses the most promising sequencing rules that should be used in pure greedy constructive heuristics. Further experiments and results are reported and discussed in the fourth section. In the last section, we draw some final conclusions.

2   The Nesting Problem and Underlying Concepts

The nesting problem addressed in this paper consists in finding the best layout for a set of 2-dimensional pieces on a board so that no pieces overlap and all of them are placed inside the board. The pieces have irregular contours that may potentially include concavities. The board is a strip with a fixed width and a length virtually unlimited. The quality of a layout is measured as the length of the used strip, or equivalently, as the usage of the area occupied up to the length it reaches on the strip. Due to technological constraints and physical properties of the raw materials, pieces are only allowed to be placed under a discrete set of orientations. Furthermore, the pieces may be placed in any unused position of the strip given that the internal part of both the pieces and the strip are homogeneous.

The geometrical representation of the pieces has a significant impact on the efficiency of the solution approaches. Pieces can be represented by defining their vertices as sequences of points. Although this is a simple representation, it requires complex trigonometric operations to compute the relative position of the shapes. An alternative is based on a raster representation (grid) where shapes are represented through a matrix of values (pixels). This representation can be used to check easily for overlaps, in particular when it is combined with the use of no-fit polygons (NFP). However, the accuracy of the representation depends on the unit size of the grid, and it increases, as the unit size gets smaller. Similarly, the computational burden increases significantly with smaller unit sizes, due to the large number of total grid units used. The raster representation is also more adequate for pieces with orthogonal edges, since non-orthogonal edges are only approximated.

The placement of the pieces is usually tackled with the assistance of NFP and inner-fit polygons (IFP) [1]. The NFP between two pieces can be described as a set of points that define the relative position between two polygons (whether they are overlapping or touching). Its main advantage is the simplification of the overlap verification process, but it is only efficient for discrete orientations since they can be computed offline in a pre-processing stage. The IFP is similar to the NFP, but it is used to ensure that a polygon is placed inside another.

Given the computational burden involved in the geometric operations inherent to the nesting problem, a promising approach has been to consider the use of dedicated hardware to support this computation, and in particular, the GPU [6]. The GPU is able to execute operations on multiple pixels simultaneously, which can be explored to improve the efficiency of the approaches when compared to the use of a normal CPU. The rasterization capabilities of the GPU enable producing feasible layouts considering the raster representation of both the pieces and the board. Using a higher representation quality also requires an increase in the total number of pixels, leading to a higher computational cost.

3 Solution Framework

In this paper, we explore and analyze a GPU-based greedy heuristic for the nesting problem. To take full advantage of the GPU, we used a raster representation to define the pieces and layout. The heuristic relies on a left-bottom placement rule that places the pieces iteratively in the layout according to a predefined sequence. Given the importance of these sequences on the quality of the layouts, we study the influence of these sequences on the performance of the heuristic.

Sofia et al. [7] present details about the GPU-based placement heuristic used in this work (GPU-Nest). Only a brief description of this heuristic is presented here. Given a predefined sequence of pieces, the heuristic places the pieces iteratively using a left-bottom placement rule. At each iteration, the placement rule is used to determine the most left-bottom feasible placement point of the next piece for all admissible orientations. Among all the alternatives, the heuristic selects the one that corresponds to the leftmost and bottommost position. In our implementation, this is achieved by keeping buffers with the admissible placement positions (pixels), one buffer for each piece type and orientation. Each time one piece is placed in the layout, the buffers are updated by drawing one NFP in each one (the NFP between the placed piece and the buffer piece type and orientation). This process has a drawback since it is not able to detect perfect fit situations due to the rasterization. To overcome this issue, the pieces are placed with a gap of 1 pixel between them. The heuristic was completely implemented in the GPU, which makes it computationally efficient. The NFPs and IFPs of all the pieces are generated in a pre-processing phase, and copied to the GPU memory.

The main control parameters are the grid resolution (which impacts on the approximation quality of the pieces), and the empty space left between pieces (in order to allow placement of pieces in small holes). An increase in the grid resolution leads to better approximations, which may reduce the layout length and the empty spaces between the pieces, while it also increases the computational cost.

Sofia *et al* [7] explored several greedy criteria to create static sequences of pieces. The static criteria implemented were the following: random (pieces are sorted randomly), height (pieces are ordered by height, taller first), width (pieces are ordered by width, wider first), irregularity (pieces are ordered by irregularity, most irregular first), rectangularity (pieces are ordered by rectangularity, less rectangular first), and size (pieces are ordered by area, larger first). The results clearly showed a significant influence of the criteria on the layout quality, with the best results being obtained by the size criteria. Several other authors obtained similar results. One of the main reasons for this conclusion is due to the fact that the layout length is mainly determined by the placement of the large pieces, since the smallest ones can frequently be accommodated among the largest ones.

The evaluation and assessment of the effectiveness of a piece sequence sorting criterion is done by comparing it to the solutions obtained by random sequences. For a given sorting criterion to be clearly effective, it should produce better layouts than the ones produced with random sequences. Additionally, the best results produced by random pieces sequences may allow identifying specific patterns that can enable the creation of new rules that consistently produce high quality layouts.

4   Computational Experiments

The computational experiments were executed on a platform with an Intel Xeon E5-5690@3.46Ghz processor, 48Gb RAM@1.33Ghz, Windows 7 x86-64, and a GPU Tesla C2070. The datasets used in the computational experiments were obtained from the ESICUP (EURO Special Interest Group on Cutting and Packing, www.fe.up.pt/esicup) website. These datasets, ordered by increasing geometric complexity, have been selected due to their diverse geometry (convex/irregular), total pieces, piece type, variation in size, and other factors (Table 1). This allows the evaluation of the approach under different circumstances. All pieces can be placed in 0 and 180 degrees rotations. Table 1 also presents the average computational times of the pre-processing phase (NFP) and the GPU-Nest heuristic (GPU-Nest). These computational times are independent of the pieces sequence, since they mainly depend on the pieces geometric characteristics. Namely, the pre-processing time depends on the number of vertices, while the nesting time depends on the total number of pieces, piece types and the number of pixels (given by the ratio between the board width and the grid resolution).

Table 1 Datasets characteristics and average computing times

Datasets	Characteristics						Average time	
	total pieces	piece types	average vertices	board width	unit grid size	total pixels (10^3)	NFP (s)	GPU-nest (s)
shapes	43	4	8.75	40	0.20	140.0	0,4	0.3
shirts	99	8	6.63	40	0.20	200.0	0.9	1.3
trousers	64	16	5.06	79	0.25	632.0	2.5	7.8
swim	48	10	21.90	5752	20.00	200.9	22.5	1.9

4.1 Randomly Generated Sequences

In order to assess the impact that a sequence may have on the final layout quality, a set of 200 randomly generated sequences were created and evaluated for the four datasets. Figure 1 shows the layout density histogram for dataset shapes. It can be seen that there is a large concentration of results within a certain range of solution quality, and that as the quality of the solutions increases or decreases, the total number of solutions produced diminishes significantly. This shows that there is room for improvement in the quality of the solutions by using better piece sequencing rules. Histograms for the other datasets (*shapes*, *swim* and *trousers*) have similar profiles. These results will serve as a baseline comparison against the results obtained with other pieces sequence sorting criteria.

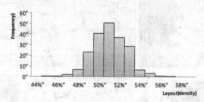

Fig. 1 Layout density histogram for dataset shapes

4.2 Divide Pieces by Size in Two Groups

An alternative to the random sequence approach was explored, based on the division of the pieces in two groups according to their area. A random sequence is generated within each group. A full sequence is created by adding the sequence from the group with the smaller pieces to the end of the sequence from the group with the bigger pieces. This strategy is based on the idea that the larger pieces define the main structure of a layout, and the smaller pieces may be placed in holes between the largest ones.

In order to test this approach based on subgroups organized by piece size, 200 sequences were produced for each subgroup in each dataset. The relative sizes of the pieces from each dataset were compared and a good allocation of pieces to each group selected. Figure 2 shows the layout density frequency polygons for datasets *shapes* and *shirts* when considering one and two groups. The two datasets clearly exhibit distinct behaviours when considering pieces divided in two groups: dataset *shapes* shows a small density improvement, while dataset *shirts* shows a bigger improvement. Dataset *swim* shows a similar behaviour to dataset *shapes*, while dataset *trousers* shows a similar behaviour to dataset *shirts*.

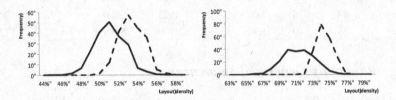

Fig. 2 Layout density frequency polygons for 1 group (solid line) and 2 groups (dashed line) for datasets *shapes* (left) and *shirts* (right)

The main reason for these different behaviours is due to the pieces' relative size in each dataset, as seen in figure 3. For dataset *shapes*, none of the smaller pieces could be placed among the bigger ones and only some of them for dataset *swim*. On the other hand, for datasets *trousers* and *shirts* the smaller pieces were all placed in holes between the larger ones, without increasing the layout length.

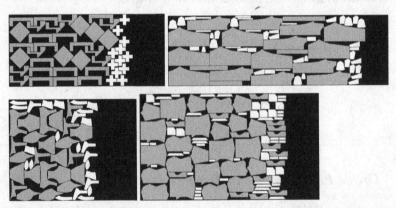

Fig. 3 Layouts with 2 groups of pieces (largest pieces in grey, smallest in white) for datasets *shapes* (top-left), *trousers* (top-right), *swim* (bottom-left) and *shirts* (bottom-right).

4.3   Best of 20 Random Sequences Divided in Two Groups

The solution quality obtained when using only one random sequence divided by groups has a large variability (figure 2). To overcome this issue and achieve a trade-off between the solution quality and the computational time, we tested an alternative approach where 20 random sequences, divided in two groups, are created and the best solution is selected. Table 2 compares the results between ordering the pieces by size (static criteria, i.e., fixed sequence), and random sequences (one and two groups). In the last column, we provide the results when considering the best layout among 20 random sequences generated with two groups. Selecting the best of 20 layouts allows obtaining a good solution, which has a very high probability of being better than the average solution of 200 random sequences. The downside of this approach is the increased computational cost.

Table 2 Impact of sequence generation procedure in the solution quality

Dataset	(1 seq,) static	(200 random seq.) 1 group		(200 random seq.) 2 groups		(20 random seq.) 2 groups
	—	avg.	(st.dev.)	avg.	(st.dev.)	best
shapes	51.7%	51.0%	(1.6%)	53.3%	(1.3%)	53.8%
shirts	74.8%	71.2%	(1.8%)	74.3%	(0.9%)	76.3%
trousers	76.5%	71.1%	(2.4%)	76.6%	(2.1%)	78.5%
swim	58.4%	56.2%	(1.8%)	59.3%	(1.6%)	58.9%

These algorithms have been implemented in the GPU, which improves the computation efficiency of this approach. It is specially true for datasets with complex geometries, as shown by the GPU-Nest heuristic average times on table 1. For instance, the best of 20 layouts for dataset *swim*, where pieces have in average more than 20 vertices, can be obtained in about 1 minute.

5   Conclusion

The results reported in this paper show that the solution quality of the layout is strongly dependent on the combination of the pieces sequence and placement rule. These results enabled the setting of a baseline for comparison for other experiments. The definition of the sequence based on the division into ordered sets of pieces allowed to determine the influence that such sequences may have on the quality of the layouts with a certain degree of variability. Noticeable improvements over the base random sequences were achieved with a simple division into two sets. The datasets with the largest variation in piece size showed to benefit significantly from this sequencing method, while more uniform datasets did not. Further research may lead to sequencing rules that enable generating consistently high quality solutions. Due to the implementation of these algorithms in a GPU, significant improvements in computational efficiency were achieved.

Acknowledgments. This work is financed by the FCT -- Fundação para a Ciência e a Tecnologia (Portuguese Foundation for Science and Technology) within project UID/EEA/50014/2013.

References

[1] Bennell, J., Oliveira, J.: The geometry of nesting problems: A tutorial. European Journal of Operational Research 184, 397–415 (2008)

[2] Bennell, J., Oliveira, J.: A tutorial in irregular shape packing problems. Journal of the Operational Research Society 60, 93–105 (2009)

[3] Burke, E., Hellier, R., Kendall, G., Whitwell, G.: A new bottom-left-fill heuristic algorithm for the two-dimensional irregular packing problem. Operations Research 54, 587–601 (2006)

[4] Fowler, R.J., Paterson, M.S., Tanimoto, S.L.: Optimal packing and covering in the plane are np-complete. Information Processing Letters 12, 133–137 (1981)

[5] Gomes, A.M., Oliveira, J.: A 2-exchange heuristic for nesting problems. European Journal of Operational Research 141, 359–370 (2002)

[6] Owens, J.D., Luebke, L., Govindaraju, N., Harris, M., Krüger, J., Lefohn, A.E., Purcell, T.J.: A survey of general-purpose computation on graphics hardware. Computer Graphics Forum 26, 1467–8659 (2007)

[7] Sampaio, S., Gomes, A.M., Rodrigues, R.: Exploring graphical processing in irregular strip packing problems. To be published in CIM Series in Mathematical Sciences, by Springer Verlag, for IO2013 - XVI Congresso da APDIO (June 2013)

[8] Wäscher, G., Haußner, H., Schumann, H.: An improved typology of cutting and packing problems. European Journal of Operational Research 183, 1109–1130 (2007)

Measures in Sectorization Problems

Ana Maria Rodrigues[1] and José Soeiro Ferreira[2]

[1] INESC TEC and ISCAP – Instituto Politécnico do Porto, Porto, Portugal
amr@inesctec.pt
[2] INESC TEC and Faculdade de Engenharia, Universidade do Porto, Porto, Portugal
jsf@fe.up.pt

Abstract. Sectorization means dividing a whole into parts (sectors), a procedure that occurs in many contexts and applications, usually to achieve some goal or to facilitate an activity. The objective may be a better organization or simplification of a large problem into smaller sub-problems. Examples of applications are political districting and sales territory division. When designing/comparing sectors some characteristics such as contiguity, equilibrium and compactness are usually considered. This paper presents and describes new generic measures and proposes a new measure, desirability, connected with the idea of preference.

1   Introduction

Sectorization problems (SP) occur in many contexts and applications in which a large region should be divided into smaller regions that meet specific conditions such as, for example, the division of a region into political districts ([Hes65], [Ric13]) or the definition of sales territories ([Gon11], [Hes71], [Kal05]). The division of a geographical area into smaller regions facilitates routing in the collection of municipal solid waste ([Rod15a], [Mou09]) or snow removal ([Muy02], [Sal12]). Sectorization may also appear in the description of regions for health care ([Ben13]), emergency accident ([Ber77]), for policing ([Dam02], [Lun12]) and location of schools ([Tak03], [Car04]), electricity distribution areas ([Ber03]) and maintenance operations [Per08]. The detection of communities in social networks is another recent application: finding groups of individuals (vertices) within which connections are dense but sparser between them ([New04]). Clustering and Sectorization present different motivations ([Kal05]): the first aims the dissimilarity between groups and the second (in territorial design, for instance) has just the opposite intention. Being the applications of SP so abundant and important, their study becomes of great relevance to society and also to science, since SP do not have an easy solution, and the solution methods used are highly dependent on applications, as can be attested by the above-mentioned references. These methods

© Springer International Publishing Switzerland 2015
A.P.F.D. Barbosa Póvoa and J.L. de Miranda (eds.), *Operations Research and Big Data*,
Studies in Big Data 15, DOI: 10.1007/978-3-319-24154-8_24

range from exact, using mathematical programming, to heuristic approaches, in a single or multi-criteria framework. But even those that follow "exact" models often end up in approximate methods.

Commonly, there are a set of common features to be preserved in the construction or evaluation of sectors, such as: Equilibrium (identical proportion of the parts in relation to the whole), Compactness (sectors showing regular shapes, circles or squares, avoiding "tentacles" ...) and Contiguity (elements of each sector arranged in single "body", thus avoiding breaking the sector into small portions). These three characteristics are not unique, although they account for most relevant applications. Therefore, the evaluation of the quality of the sectors, obtained by any method, is a difficult task, due to the differently defined metrics involved.

This paper discusses and proposes some general and new measures that can be used to assess the quality of sectors and subsequent comparison of the results obtained. In a more advanced stage of the ongoing research work, the authors intend to integrate these general measures with a sectorization approach based on an analogy with electromagnetism forces [Rod15b] and with a multi-criteria method described in [Fer13]. That may result in a "universal method" for many real sectorization problems, if appropriate adjustments are provided.

2   Measures

The most common measures or criteria used in Sectorization contemplate, in some way, conceptions or definitions of Equilibrium, Compactness and Contiguity. Other criteria such as integrity, in political districting, or racial balance in school districting, are used in specific applications. In this paper, in addition to advancing general definitions for the most common measures, another one, related with desirability, is proposed.

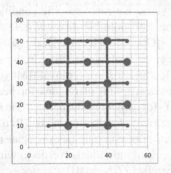

The authors consider that the definitions should be simple, transparent and clear to decision makers and easily adapted to cover most current real situations. It may be questionable the use of complex measures when, often, the models need to be simplified and/or decisions must be taken in the absence of clear facts or data.

Fig. 1 Group with 25 elementary units to construct 4 sectors

For the sake of illustration of the measures, a small example with 25 elementary units to be aggregated into 4 sectors is presented in Fig. 1. Suppose that a quantity is assigned to each point, in this case a quantity 1 assigned to a small grey circle and a quantity 2 assigned to a large grey circle. The total quantity equals 37 (12 large circles plus 13 small circles). Lines between circles establish relations between points.

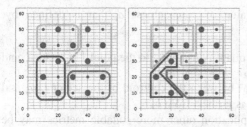

Fig. 2 Two sectorizations named Left and Right

Suppose that each of the four sectors has a limit quantity of 10, enough to accommodate the quantity 37, and that after applying different methods, two distinct sectorizations were obtained, "Left" and "Right" (see Fig. 2).

The example presented in Fig.1 with the initial map and the two proposed groups of sectors in Fig. 2 are used to illustrate the different measures.

2.1 Equilibrium

To evaluate the equilibrium, that is, the similarity between the quantities (number of electors, amount of work, quantity of waste to collect,...) in each sector, the proposed measure is the coefficient of variation (CV_q). It is calculated as follows, for a group of k sectors with quantities q_i, $i=1,...,k$:

$$CV_q = \frac{s'_q}{\overline{q}} \text{ , where } \overline{q} = \frac{\sum_{i=1}^{k} q_i}{k} \text{ and } s'_q = \sqrt{\frac{1}{k-1} \sum_{i=1}^{k}(q_i - \overline{q})^2} \text{ .}$$

Balanced sectors should have a CV_q as close to zero as possible. The use of the coefficient of variation allows the comparison between sectors obtained with different original sets.

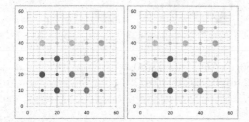

If only the equilibrium is considered, the Left solution is better than the Right, $CV_q(Left)$ $<CV_q(Right)$).

Fig. 3 Two solutions – Left: $CV_q = 0.054$ and Right: $CV_q = 0.162$

2.2 Compactness

Let us associate the concept of *compactness* to the idea of *concentration* or *density*. Higher concentration should avoid sparse sectors.

Each sector i ($i=1,...,k$) has a value of compactness d_i that is defined by:

$$d_i = \frac{\sum_j q_{ij}}{dist(o_i, p_i)}, \quad q_{ij} \text{ represents the quantity assigned to the point } j \text{ in sector } i, \text{ and}$$

$dist(o_i, p_i)$ is the distance (Euclidean) between the centroid of the sector i, o_i, and the point of the same sector, p_i, that is farthest from o_i.

Depending on the particular application other more suitable metrics may be considered. In fact, in a broad sense, compactness may be defined as the total quantity (in a sector) divided by the number of elements of that sector.

Higher values of d_i represent higher values of compactness, which means a "higher density" in sector i. It is desirable that, regarding the same *map*, different sectors appear with similar values of compactness. To quantify that similarity it is proposed the coefficient of variation of compactness given by CV_d: $CV_d = \frac{s'_d}{\bar{d}}$,

where $\bar{d} = \frac{\sum_{i=1}^{k} d_i}{k}$ and $s'_q = \sqrt{\frac{1}{k-1}\sum_{i=1}^{k}(q_i - \bar{q})^2}$. A good sectorization must have a CV_d close to zero.

Once again the example is revisited and showed in Fig.4.

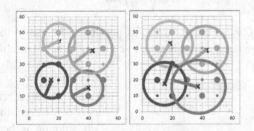

Regarding compactness, the Right solution is better than the Left (CV_d(Right)> CV_d(Left)).

Fig. 4 Left CV_d=0.7606 (d_1=0.8050; d_2=0.6472; d_3=0.8050; d_4=0.8050) and Right CV_d=0.1192 (d1=0.5824; d_2=0.6472; d_3=0.6685; d_4=0.5093)

2.3 Contiguity

The contiguity of sectors is another recurrent and important feature to considerer when evaluating the quality of sectors. Depending on the application it is desirable that each sector form just one body, or more *strongly*, that the interceptions between sectors are as small as possible. Two types are considered: Strong Contiguity and Weak Contiguity.

Strong Contiguity - A sectorization presents strong contiguity if the area of each smaller convex polygon, containing all elements of each sector, does not overlap the others. This condition may be even stronger: the areas should not touch each other. Only the Left solution shown in Fig.5 has absolute strong contiguity.

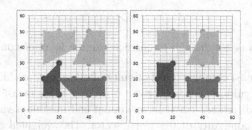

Fig. 5 *Left* sectors with strong contiguity; in *Right* sectors, red and blue touch each other

Weak Contiguity - is concerned with the shape of the sectors, and with the links between elementary units within each sector. If the subgraph induced by the elements (vertices) which represent the sector and links (arcs and edges) is connected, that is, there is a walk between any pair of elements of the sector without leaving it, the sector presents weak contiguity.

The evaluation of the weak contiguity of the k sectors is calculated using the adjacency matrices obtained from the k subgraphs $G_i' = (V'_i, E'_i)$ $(i=1,...,k)$, where V'_i and E_i represent the set of vertices and the set of edges of subgraph G'_i, respectively. The number of vertices of each sector i is represented by: $|V_i|=n_i$, $i=1,...,k$. For each subgraph, G'_i also considers the symmetric matrix given by $M^i=[m^i_{wj}]_{w,j=1,...,ni}$ with principal diagonal with zeros

$$M^i = \begin{bmatrix} 0 & m^i_{12} & m^i_{13} & \cdots & m^i_{1ni} \\ m^i_{21} & 0 & m^i_{23} & \cdots & m^i_{2n_i} \\ \vdots & \vdots & \vdots & \ddots & \vdots \\ m^i_{n_i1} & m^i_{n_i2} & m^i_{n_i3} & \cdots & 0 \end{bmatrix}$$

where $m^i_{wj} = \begin{cases} 1, \text{ if in sector } i \text{ exists a walk between } w \text{ and } j \\ 0, \text{ otherwise} \end{cases}$

If for all $j \in \{1,...,n_i\}$, $\sum_{w=1}^{n_i} m^i_{wj} = n_i - 1$ or, which is equivalent, for all $w \in \{1,...,n_i\}$ the condition $\sum_{j=1}^{n_i} m^i_{wj} = n_i - 1$ is verified, then sector i is contiguous. But different levels of contiguity must be considered.

The next expression is used, as a measure of contiguity (c_i) for each sector i:

$$c_i = \frac{\sum_{j=1}^{n_i}\left(\sum_{w=1}^{n_i} m_{wj}^i\right)}{n_i(n_i-1)}$$. This is not enough to characterize the level of contiguity.

The quality of the sectors must combine the contiguity of all sectors produced. To evaluate the resulting contiguity, depending on the objective, some measures can be considered such as the difference between maximum and minimum values of contiguity or the value of contiguity of the worst sector. The weighted average ($\bar{c}$) of *isolated contiguities*, is proposed:

$$\bar{c} = \frac{\sum_{i=1}^{k} c_i \cdot n_i}{N}$$. $\bar{c}$ is a value that is always between 0 and 1. Using weights the final contiguity is proportional to the size of the sector.

From the perspective of contiguity, a good sectorization must have a $\bar{c}$ value as close to 1 as possible.

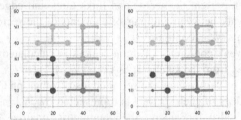

When only (weak) contiguity is consi-dered, Left is preferred.

Fig. 6 Left $\bar{c}=1$ ($c_1=1$; $c_2=1$; $c_3=1$; $c_4=1$) and Right $\bar{c}=0.7867$ ($c_1=0.71429$; $c_2=1$; $c_3=0.1667$; $c_4=1$)

2.4   Desirability

Suppose the situation in which it would be desirable that some specific groups of elements belong to the same sector. It is just a preference, and not a mandatory situation. For some reason, eventually past experiences, the decision maker prefers that some groups of points stay together: served by the same vehicle, vote in the same district or attend the same school, depending on the application.

For instance, after a sectors construction, the *degree of desirability* should measure how close the solution is from the previous preferences of the decision maker.

In this context, let F represent the number of groups identified by the decision maker as having some affinity and f_i, $i=1,...,F$, the number of elementary units in each of the predefined F groups.

If the sectorization involves the creation of k sectors, the maximum number (H) of elementary units out of the F groups, defined by the decision maker, is

$H = \sum_{i=1}^{F} \left(f_i - \min\left\{ f_i - 1; \left\lceil \dfrac{f_i}{k} \right\rceil \right\} \right)$. Then, the measure of desirability (*Des*) is de-

fined by: $Des = 1 - \dfrac{NE}{H}$, where *NE* is the number of elements out of the groups (after sectorization). In the following, we illustrate the definition by considering the example of Fig. 7, where $k=4$, $f_1=6$ and $f_2=4$.

$H = f_1 - \min\left\{ f_1 - 1; \left\lceil \dfrac{f_1}{k} \right\rceil \right\} + f2 - \min\left\{ f_2 - 1; \left\lceil \dfrac{f_2}{k} \right\rceil \right\} = 6 - \min\{5;2\} + 4 - \min\{3;1\} = 7$

This means that, in the worst case (a sectorization completely disrespecting the preferences), the number of outsiders is equal to 7.

Left and Right sectorizations have *NE* equal to 2 and 4, respectively (Fig.7).

Finally, $Des = 1 - \dfrac{2}{7}$ (Left) and $Des = 1 - \dfrac{4}{7}$ (Right). In summary, Left sectorization is closer to the previous desire of decision maker.

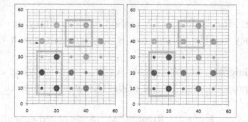

Values of *desirability* near to 1 indicate that groups selected are mostly respected by the sectorization process. On the other hand, values near to 0 indicate that the sectorization made is not respecting the prior preferences.

Fig. 7 Grey rectangles represent groups defined by the decision maker. Left *Des*= 0.7143 and Right *Des*=0.4286

If only desirability is considered, Left is better than Right.

3   Conclusion

Sectorization problems (SP) occur in many contexts and real applications, as described. Solving SP is a complex task, as mentioned in many publications, and it is usually indispensable to contemplate one or more characteristics of the sectors, to evaluate the quality of the solutions. Typical properties are connected with the ideas of equilibrium, compactness and contiguity.

This paper also considered these properties, while proposing new measures or criteria and aiming at defining them as generic and transparent as possible, so that they can be used in most practical SP. Desirability is another and new measure introduced. The idea is to embrace possible desirabilities (or preferences) of a decision maker.

This work is part of the authors' ongoing research about SP, which encompasses the integration of these generic measures with a general sectorization approach based on an analogy with Electromagnetism [Rod15b] and with a multi-criteria soft method described in [Fer13]. The final intention is to devise a "universal method" capable of dealing with many real SP, if appropriate adjustments are provided.

Acknowledgments. This work was partially financed by National Funds through the FCT-Portuguese Foundation for Science and Technology within project "Project SEROW/PTDC/EGE-GES/121406/2010" and by the North Portugal Regional Operational Programme (ON.2 - O Novo Norte), under the National Strategic Reference Framework (NSRF), through the European Regional Development Fund (ERDF), and by National Funds, through the FCT within "Project NORTE-07-0124-FEDER-000057".

References

[Ben13] Benzarti, E., Sahin, E., Dallery, Y.: Operations management applied to home care services: Analysis of the districting problem. Decision Support Systems 55, 587–598 (2013)

[Ber03] Bergey, P.K., Ragsdale, C.T., Hoskote, M.: A Simulated Annealing Genetic Algorithm for the Electrical Power Districting Problem. Annals of Operations Research 121, 33–55 (2003)

[Ber77] Bertolazzi, P., Bianco, L., Ricciardelli, S.: A method for determining the optimal districting in urban emergency services. Computers & Operations Research 4, 1–12 (1977)

[Car04] Caro, F., Shirabe, T., Guignard, M., Weintraub, A.: School redistricting: embedding GIS tools with integer programming. Journal of the Operational Research Society 55, 836–849 (2004)

[Dam02] D'Amico, S.J., Wang, S.-J., Batta, R., Rump, C.M.: A simulated annealing approach to police district design. Computers & Operations Research 29, 667–684 (2002)

[Fer13] Ferreira, J.S.: Multimethodology in Metaheuristics. Journal of the Operational Research Society 64, 873–883 (2013)

[Gon11] González-Ramírez, R., Smith, N.R., Askin, R.G., Miranda, P.A., Sánchez, J.M.: A Hybrid Metaheuristic Approach to Optimize the Districting Design of a Parcel Company. Journal of Applied Research and Technology 9, 19–35 (2011)

[Hes65] Hess, S., Weaver, J., Siegfeldt, H., Whelan, J., Zitlau, P.: Nonpartisan Political redistricting by computer. Operations Research 13, 998–1006 (1965)

[Hes71] Hess, S.W., Samuels, S.A.: Experiences with a sales districting model: criteria and implementation. Management Science 18, 41–54 (1971)

[Kal05] Kalcsics, J., Nickel, S., Schröder, M.: Towards a unified territorial design approach - applications, algorithms and GIS integration, Top, 13, 1–56. Springer (2005)

[Lun12] Lunday, B.J., Sherali, H.D., Lunday, K.E.: The coastal seaspace patrol sector design and allocation problem. Computational Management Science 9, 483–514 (2012)

[Mou09] Mourão, M.C., Nunes, A.C., Prins, C.: Heuristic methods for the sectoring arc routing problem. European Journal of Operational Research 196, 856–868 (2009)

[Muy02] Muyldermans, L., Cattrysse, D., Van Oudheusden, D., Lotan, T.: Districting for salt spreading operations. European Journal of Operational Research 139, 521–532 (2002)

[Per08] Perrier, N., Langevin, A., Campbell, J.F.: The sector design and assignment problem for snow disposal operations. European Journal of Operational Research 189, 508–525 (2008)

[Ric13] Ricca, F., Scozzari, A., Simeone, B.: Political Districting: from classical models to recent approaches. Annals of Operations Research 204, 271–299 (2013)

[Rod15a] Rodrigues, A.M., Ferreira, J.S.: Waste collection routing - limited multiple landfills and heterogeneous fleet. Networks 65, 155–165 (2015)

[Rod15b] Rodrigues, A.M., Ferreira, J.S.: Sectors and Routes in Solid Waste Collection, to be published in Springer-Verlag - CIM Series in Mathematical Sciences (2015)

[Sal12] Salazar-Aguilar, M.A., Langevin, A., Laporte, G.: Synchronized arc routing for snow plowing operations. Computers & Operations Research 39, 1432–1440 (2012)

[Tak03] Takashi, T., Yukio, S.: Evaluation of School Family System Using GIS. Geographical Review of Japan 76, 743–758 (2003)

Integrated Cutting and Production Planning: A Case Study in a Home Textile Manufacturing Company

Elsa Silva, Cátia Viães, José F. Oliveira, and Maria Antónia Carravilla

INESC TEC and Faculdade de Engenharia, Universidade do Porto, Portugal
emsilva@inescporto.pt, catia.viaes@gmail.com,
{jfo,mac}@fe.up.pt

Abstract. In this paper we consider the problem of minimizing the waste of textile material in a Portuguese home textile manufacturing company. The company has a vertical structure covering the different production stages of the home textile, from weaving until the finished products. Production planning comprises different decisions: the definition of the widths and lengths of the fabric rolls to be produced, the number of fabric rolls to be used from stock or purchased and the definition of the cutting patterns to be applied to each width of the fabric roll, so that the waste is minimized. We propose a MIP model, solved by a column generation method, to tackle the problem.

1 Introduction

The study of the production planning problem integrated with the cutting process on a home textile manufacturing company is the focus of this paper. The company is located in the north of Portugal and the final products are sheets, pillowcases and duvet covers.

The production process of this textile company has the following phases: weaving, dyeing and/or printing, cutting, sewing and packaging. The last two phases are not considered in this study since they do not contribute to the waste of textile material. Between each pair of the remaining phases there is an intermediate stock to be used in a subsequent production phase, arising either from a finished product of an upstream phase or purchased from an external supplier. The production process is illustrated in Figure 1.

In the weaving process, the fabric rolls are produced in industrial looms. The rolls are then dyed in a bathtub where they acquire the final color. In the printing phase drawings of one or more colors are reproduced on the fabric rolls through a serigraphy process. This process is optional as the customer may decide on plain or patterned products. The process ends at the cutting phase where the fabric rolls are unrolled and folded in layers on a cutting table and are cut following a cutting

A.P.F.D. Barbosa Póvoa and J.L. de Miranda (eds.), *Operations Research and Big Data*,
Studies in Big Data 15, DOI: 10.1007/978-3-319-24154-8_25

Fig. 1 Production process
(Weaving-Cutting).

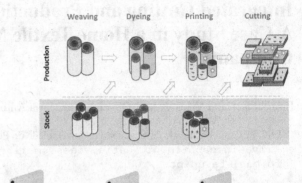

Fig. 2 Fabric roll on the
cutting table.

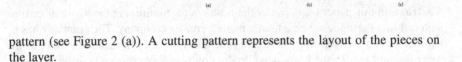

pattern (see Figure 2 (a)). A cutting pattern represents the layout of the pieces on
the layer.

2 Cutting Process

The layer is cut manually with a saw, thus the cutting patterns must be of type
guillotine. In a guillotine type cutting pattern the cuts are made along the entire
length/width of the cutting pattern and cannot stop in the middle of the path (see
Figure 3). Another important characteristic of the cutting patterns is the number of
stages, i.e. the number of cuts with successively different directions that must be
performed by the saw, in order to cut the pieces as planned in the cutting pattern.
The cutting patterns considered by the company are at most three-stage, eventually
with an additional cut performed to separate the pieces from waste, called trimming.
An example of a three-stage cutting pattern with trimming is presented in Figure 3.

The technique used to spread the fabric rolls on the cutting table enables to cut
"half-pieces" by using the folds between layers (see 2 (b)). This is however not
possible for orders in which a high accuracy in the dimensions of the pieces is
needed. In this case a cut at the fabric roll ends is required, in order to divide the
layers, producing additional waste (see 2 (c)).

The problem has firstly been addressed by Cerqueira (2013) and a mixed integer programming model was proposed, however no computational experiments
have been conducted and a further analysis has been indicated as future work. In
this study we propose to combine the model proposed in Cerqueira (2013) with an
adaptation of the heuristic proposed by Almeida (2014) for the generation of cutting
patterns.

Fig. 3 Three-stage cutting
pattern with trimming.

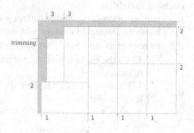

3 Integer Programming Model

During the production process, any manufactured product must be identified by a reference in accordance with its characteristics of weaving $t \in \mathscr{T}$; $\mathscr{T} = \{1,\ldots,T\}$, dyeing $c \in \mathscr{C}$; $\mathscr{C} = \{1,\ldots,C\}$ and printing $e \in \mathscr{E}$; $\mathscr{E} = \{1,\ldots,E\}$. In the weaving phase, besides the minimum production quantity (L_{min}^t) constraint, only fabric rolls with predefined widths for each weaving reference $(\mathscr{J}_t)$ are produced.

As previously defined, in each production phase intermediate stock can be used for each width j in reference tce, represented by S_j^{tce}. In the cutting process too short cutting patterns are not desirable, thus a minimum length P_{min} is defined for each cutting pattern. The cutting pattern p have a length l_p, a width w_p and is composed by a_{ip} pieces of type i and has a cutting process time (T_p). In what concerns the cutting table, there is a maximum length P_{max}, and a limit on the maximum height h_{max}^{tce}, defined by the maximum number of layers of fabric with reference tce. The ordered piece i of reference tce is characterized by a length l_i, a width w_i and a lower q_i^{LB} and upper bound demand q_i^{UB}. The following mixed integer programming model was adapted from Cerqueira (2013) the first reference to this problem.

The objective function expressed by equation (1), aims at minimizing the fabric cost and the wastage. The fabric cost is composed by the weighted sum of the cost to produce fabric throughout the production process (K_{1j}^{tce}), the cost of using fabric from *stock* (K_{2j}^{tce}) and the wastage is considered in the number of cutting tables for each reference tce. The division of (x_p) by the maximum number of layers in each cutting table h_{max}^{tce} determines the total number of cutting tables for reference tce in which the cutting pattern $\mathscr{P}$.

In constraints (2)-(3) it is ensured that the total production is at least of L_{min}^t, if there exists production of the fabric roll with weaving reference t and width j $(\beta_j^t = 1)$. Constraint (4) ensures that the quantities used from stock for the reference tce and width j effectively exist. The overall length used in the cutting process with reference tce in the width j is obtained from the production in earlier phases (y_j^{tce}) or from stock (z_j^{tce}). This constraint is ensured in (5). The minimum and maximum production for each piece i for the combination tce is ensured by constraints (6) and (7). Constraints (8) and (9) ensure that, if a cutting pattern is used, it will be applied at least to a minimum length of fabric (P_{min}), the big M considered to be equal to the sum of all the ordered pieces times the corresponding demand. The last four constraints (10)-(13) define the domains of the decision variables.

Decision variables

x_p - number of times the cutting pattern p is used, $p \in \mathscr{P}$;

y_j^{tce} - number of meters to produce with width j and reference tce, $j \in \mathscr{J}_t, t \in \mathscr{T}, c \in \mathscr{C}, e \in \mathscr{E}$;

z_j^{tce} - number of meters to use from stock with width j and reference tce, $j \in \mathscr{J}_t, t \in \mathscr{T}, c \in \mathscr{C}, e \in \mathscr{E}$;

$\beta_j^t = \begin{cases} 1, & \text{if exist producion of the reference } t \text{ with width } j, t \in \mathscr{T}, j \in \mathscr{J}_t \\ 0, & \text{otherwise} \end{cases}$

$\alpha_p = \begin{cases} 1, & \text{if cutting pattern } p \text{ is used}, p \in \mathscr{P} \\ 0, & \text{otherwise} \end{cases}$

$$\text{Minimize} \quad \sum_{jtce} K_{1j}^{tce} y_j^{tce} + \sum_{jtce} K_{2j}^{tce} z_j^{tce} + \sum_p \left\lceil \frac{x_p}{h_{max}^{tce}} \right\rceil \tag{1}$$

$$\text{Subject to:} \quad \sum_{c \in \mathscr{C}, e \in \mathscr{E}} y_j^{tce} \geq L_{min}^t \cdot \beta_j^t, \qquad \forall t \in \mathscr{T}, j \in \mathscr{J}_t; \tag{2}$$

$$\sum_{c \in \mathscr{C}, e \in \mathscr{E}} y_j^{tce} \leq M \cdot \beta_j^t, \qquad \forall t \in \mathscr{T}, j \in \mathscr{J}_t; \tag{3}$$

$$z_j^{tce} \leq S_j^{tce}, \qquad \forall j \in \mathscr{J}_t, t \in \mathscr{T}, c \in \mathscr{C}, e \in \mathscr{E}; \tag{4}$$

$$\sum_{p:jtce} L_p x_p - y_j^{tce} - z_j^{tce} \leq 0, \qquad \forall j \in \mathscr{J}_t, t \in \mathscr{T}, c \in \mathscr{C}, e \in \mathscr{E}; \tag{5}$$

$$\sum_{p \in \mathscr{P}} a_{ip} x_p \geq q_i^{LB}, \qquad \forall i \in \mathscr{I}; \tag{6}$$

$$\sum_{p \in \mathscr{P}} a_{ip} x_p \leq q_i^{UB}, \qquad \forall i \in \mathscr{I}; \tag{7}$$

$$L_p x_p - P_{min} \alpha_p \geq 0, \qquad \forall p \in \mathscr{P}; \tag{8}$$

$$x_p - M \alpha_p \leq 0, \qquad \forall p \in \mathscr{P}; \tag{9}$$

$$x_p \geq 0 \text{ and integer}, \qquad \forall p \in \mathscr{P}; \tag{10}$$

$$\beta_j^t \in \{0,1\} \qquad \forall j \in \mathscr{J}_t, t \in \mathscr{T}; \tag{11}$$

$$y_j^{tce} \geq 0, \quad z_j^{tce} \geq 0, \qquad \forall j \in \mathscr{J}_t, t \in \mathscr{T}, c \in \mathscr{C}, e \in \mathscr{E}; \tag{12}$$

$$\alpha_p \in \{0,1\}, \qquad \forall p \in \mathscr{P}. \tag{13}$$

The model presented determines optimally the length of fabric with reference tce and width j to be produced and/or to be used from stock, as well as the selection of the cutting patterns that generate the best solution, determining how many times the cutting patterns should be used.

4 Solution Approach

It is assumed, in the Mixed Integer Programming Model (1)-(13), that the complete set P of cutting patterns is known. However, an explicit search of P may be computationally impossible when $|P|$ is huge. In practice, one starts by considering the linear relaxation of (1)-(13) and works with a reasonable subset $\bar{P} \subseteq P$ of columns, with a Restricted Master Problem (RMP). The linear relaxation of the RMP is solved via column generation, where the pricing subproblem is a *two-dimensional knapsack problem*.

Attractive columns are added to the RMP derived from the pricing subproblem solution, where a new column represents a new cutting pattern and its corresponding length. The pricing subproblem uses the dual values (Π) associated with constraints (7), ensuring the minimum production of piece i. The geometrical part of the problem is considered in the pricing subproblem, where it is ensured that the pieces completely lie on the cutting pattern, do not overlap and the cutting patterns are of type guillotine.

The method adopted is based on the delayed column generation approach proposed by Oliveira & Ferreira (1994). A heuristic is used to generate cutting patterns for each reference *tce* and width j and, when the cutting patterns heuristically obtained do not improve the current value of the RMP, the pricing subproblem is solved to optimality for 2-stage cutting patterns (with trimming) and the length of the cutting table is set to the pattern length. The column generation process starts with a set of columns representative of each piece type i with reference *tce* and width j. The initial cutting patterns are composed by one piece of type i with reference *tce* and the cutting pattern length is equal to the maximum length of the cutting table (P_{max}). The initial columns/patterns often considered the maximum number of pieces that fit in the cutting pattern, however, in order to avoid infeasible solutions due to the upper bound constraint, we consider only one-piece patterns.

The role of the pricing subproblem is to provide cutting patterns that price out profitably or to prove that none exists and a separate pricing subproblem is considered for each reference *tce* and width j. Firstly the greedy heuristic proposed by Almeida (2014) is used. A priority criterion is used for piece selection: the pieces with the highest dual value and highest demand are firstly chosen. The heuristic allows the generation of cutting patterns of type two-stage with trimming and three-stage, with and without trimming and with "half-pieces". In each iteration of the column generation process the current value of the linear relaxation of the RMP (Z_{RMP}^{lr}) is updated.

When the heuristic does not improve the value of (Z_{RMP}^{lr}), the pricing subproblem is solved by the exact method proposed by Gilmore & Gomory (1965). The pricing subproblem is now composed by a sequence of two knapsack problem types, allowing the generation of cutting patterns of the type two-stage with trimming. The first knapsack problem type creates stripes, while the second knapsack aggregates the horizontal stripes generating a cutting pattern. A total of $n + 1$ knapsack problems are solved, it should be noted that this method of solving a set of knapsack problems will be applied as often as the references *tce* and with width j. The column generation process ends when the value of (Z_{RMP}^{lr}) is no longer improved.

As the integrality of the variables of the initial formulation (1)-(13) was relaxed and the linear relaxation was solved via column generation, the solution obtained in the end of the column generation process has values that may not be integer. This issue is addressed by defining the decision variable x_p as integer and β_j^t and α_p as binaries and solving again the problem. Thus, a final integer solution is obtained. The overall solution approach is summarized in Algorithm 1.

Algorithm 1. Algorithm of the Solution Approach

---

begin
    Define the RMP;
    Relax Integrality constraints;
    Define the initial cutting patterns of the RMP;
    Solve the RMP and obtain the dual values (Π);
    while *Solution of the RMP* (Z^{rl}_{PMR}) *is improved* **do**
        For each j,tce:
        - generate cutting patterns heuristically;
        - add the cutting patterns in the RMP (new columns);
        - Solve the RMP and obtain the dual values (Π);
    end
    while *Solution of the RMP* (Z^{rl}_{PMR}) *is improved* **do**
        For each j,tce:
        - solve a knapsack to create stripes for the different pieces i, considering (Π)
        as coefficient in the objective function;
        - generate cutting patterns by solving a knapsack that combines the stripes;
        - add the cutting patterns in the RMP (new columns);
        - Solve the RMP and obtain the dual values (Π);
    end
    Solve the Integer Programming Model;
end

---

5 Computational Results

Computational experiments have been conducted on a real instance from the home textile manufacturing company. The demand is of 18 piece types and 4 different references *tce*. This demand may be cut from 20 different fabric roll types. The main objective of the computational experiments was to evaluate the sensibility of the heuristic by considering different parameters. The solution approach has been implemented in C++ in Microsoft Visual Studio with the solver IBM ILOG Cplex 12.6. The computational experiments were performed in an Intel Core i7 with 4GB of RAM. Given the uncertainty around some of the required parameters, that not even the home textile manufacturing company was able to specify, a sensitivity analysis was run, with those parameters set for a base scenario and varied during the computational experiments.

The cost of using a fabric roll from stock (K^{tce}_{1j}) has been defined as twice the cost of producing the fabric roll (K^{tce}_{2j}). The maximum number of layers of fabric on the cutting table (h^{tce}_{max}) has been set to 20. The fabric rolls in stock have been fixed to zero for all references *tce* and widths j. The minimum length of fabric roll to produce in the weaving phase has been defined as 10% of the total length of the pieces. The minimum length at which a cutting pattern can be applied (P_{min}) has been defined as zero and the maximum length of the cutting table (P_{max}) is set to 1500 cm, which corresponds to its real length. The parameters h^{tce}_{max} and P_{max}

Table 1 Computational results for the weighted objective function

Scenario	SP_{heur}		SP_{heur}^{IP}		Final Solution						
	Z_{RMP}^{lr}	n_{col}	Z_{RMP}^{lr}	n_{col}	Z_{RMP}^{mip}	gap_{mip}	$\sum y$	$\sum z$	N_p	Avg_{np}	T_{total}
Base	518031	200	517396	120	521029	0.7	521012	0	24	14.5	64
1	518032	200	517396	120	521347	0.8	521330	0	17	20.5	15
2	518063	200	517462	80	520915	0.7	520898	0	28	12.5	71
3	518063	200	517396	120	540336	4.3	540319	0	20	17.5	64
4	518063	200	517396	120	521029	0.7	517360	3652	24	14.5	64
5	1024159	200	1023123	120	1037476	1.4	516447	4564	23	15.2	65

are kept constant in the different scenarios. In Table 1 the 6 tested scenarios are presented.

The computational results are summarized in Table 2. The solutions obtained during the column generation process using only the heuristic to solve the subproblem are presented in column SP_{heur}. Column SP_{heur}^{IP} presents the solutions obtained for the linear relaxation of the RMP after the exact method has been used for solving the subproblem. The number of columns generated by the subproblems for each method is represented by n_{col}, the integer solution is represented by Z_{RMP}^{mip}, gap_{mip} relates the value of RMP relaxed with the integer solution, $\sum y$ is the total length to be produced in centimetres, $\sum z$ is the total length in centimeters of fabric to be used from stock, N_p represents the number of different cutting patterns used in the solution and Avg_{np} is the average number of times the cutting pattern is used.

The first analysis comprises the P_{min} parameter, that is equal to zero in the base scenario and in scenario 1 is equal to 6000 (four times the length of the cutting table). An increase of the minimum length that a cutting pattern can be applied means an increase of the frequency of utilization of the cutting pattern, thus reducing the setup time, but increasing the waste. This conclusion is corroborated by the value of the objective function that increased with the increase of P_{min}.

In scenarios 2 and 3 L_{min}^t was fixed to zero and to 20% of the total length of the pieces, respectively. The increase of the minimum length of production in the weaving phase also increases the waste, since the geometric combinations are reduced. It is also interesting to notice that an increase on the value of L_{min}^t reduces the number of different widths that are used for each reference tce. In scenarios 4 and 5 the fabric roll stock was set to 913 for each type of fabric roll. This value was obtained by weighing 5% of the needs of each reference t. In this new instance the model can choose between production and stock. The coefficients of the objective function were also changed in order to analyze the behaviour of the model. As expected the solution obtained uses more fabric rolls from stock when the costs are smaller.

6 Conclusions

This paper addresses the optimization of the production planning problem integrated with the cutting problem in a home textile manufacturing company. The complete production process was analyzed and a solution approach is proposed based on the column generation technique.

The methodology developed is able to decide the quantity of fabric that should be produced in each production phase, from each reference, the quantity of fabric that should be used from stock, the respective cutting patterns that should be applied to obtain the pieces and the corresponding quantities.

Preliminary computational experiments have been conducted on a real instance and an analysis of the sensibility of the proposed approach to different parameters tuning was performed, one can conclude that the proposed heuristic is highly dependent on the different costs considered in the objective function.

Acknowledgements. The first author is financed by the FCT – Fundação para a Ciência e a Tecnologia (Portuguese Foundation for Science and Technology) within project UID/EEA/ 50014/2013 and within the grant SFRH/ BPD/98981/2013.

References

Almeida, R.N.: Redução de desperdícios no processo de corte em empresas Têxtil-Lar. Master. thesis, FEUP (2014)

Cerqueira, B.: Kaizen na indústria têxtil - Uma abordagem ao aumento de produtividade e redução de desperdício. Master thesis, FEUP (2013)

Gilmore, P.C., Gomory, R.E.: Multistage cutting stock problems of two and more dimensions. Operations Research 13(1), 94–120 (1965)

Oliveira, J.F., Ferreira, J.S.: A faster variant of the Gilmore and Gomory technique for cutting stock problems. Jorbel 34(1), 23–38 (1994)

The Cutting Stock Problem: A Case Study in a Manufacturer of Pet Vivaria

Carla Sousa[1], Elsa Silva[2], Manuel Lopes[3], and António Ramos[3]

[1] School of Engineering, Polytechnic of Porto, Portugal
   1000680@isep.ipp.pt
[2] INESC TEC, Portugal
   emsilva@inescporto.pt
[3] CIDEM, School of Engineering, Polytechnic of Porto, Portugal
   {mpl,agr}@isep.ipp.pt

Abstract. This paper addresses the problem of determining the cutting patterns of metal sheets, which arises in a manufacturer of metal cages, in order to minimize the waste, the number of cuts performed, the number of metal sheets used or a weighted combination of the three. A two stage approach, to solve a 2D guillotine cutting stock problem with single and multiple stock sizes, is presented and compared with the company approach and state-of-the-art algorithms. The results show great improvement compared to the company approach and a very good performance compared to state-of-the-art algorithms.

1 Introduction

The cutting stock problem is a combinatorial optimization problem, which belongs to the wider combinatorial optimization class of Cutting and Packing (C&P) problems. The essential form of C&P problems can be summarised as follows: given a set of large objects and a set of small items (both defined in a number of geometric dimensions), the small items must be assigned to the large objects and a dimensional objective function is optimised satisfying two geometric conditions, that is, all small items lie entirely within the large objects and the small items do not overlap Wäscher *et al.* (2007).

The relevance of this problem has been emphasized by the frequency with which it is addressed in the field of operations research and operations management. These problems appear in a wide range of real life situations. The packing of boxes into containers to minimize transport costs, the packing of boxes into pallets to minimize storage costs or the cutting of wood plates to minimize waste are just a few examples of C&P problems.

In this paper we address a cutting stock problem found in a small Portuguese company, from the northern region, that produces metal cages for pets. During the manufacturing process, large metal sheets must be cut in smaller pieces to produce the main metallic components of the vivaria. Working in an extremely competi-

© Springer International Publishing Switzerland 2015   221
A.P.F.D. Barbosa Póvoa and J.L. de Miranda (eds.), *Operations Research and Big Data*,
Studies in Big Data 15, DOI: 10.1007/978-3-319-24154-8_26

Fig. 1 Vivarium

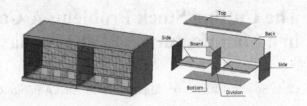

tive market, the company needs to adopt strategies to control the costs. This can be achieved, for example, by reducing the waste of raw material or by reducing the costs with inventory. The focus of this work was to develop a decision support tool, able to guide the decision making in the cutting process and also, that would not involve a new investment by the company in an optimization solver. The proposed method is based on a two-phase algorithm, firstly a set of feasible cutting patterns is defined and secondly the classical linear programming model, proposed by Gilmore & Gomory (1961), is used to decide the cutting patterns according to different objectives.

The paper is organized as follows: in the next section the problem is described. In section 3 the solution approach proposed is described. In section 4 the results are benchmarked and finally, conclusions are drawn in section 5.

2 Sheet Metal Cutting Process

The company considered in this case study produces products for pet shops, garden cages and vivaria for animal breeding and display. The research focused on the cutting division, specifically on the cutting strategies of the vivarium with the greatest demand, representing circa 1500 units in annual sales. Fig. 1 represents the vivarium and its main metallic components, which are cut from the stock metal sheets.

The raw material used for the production of the vivaria is cold rolled steel sheets, with rectangular shape and with 0.5 mm of thickness. The supplier delivers the metal sheets in three different sizes: 3000×1500 mm^2, 2500×1250 mm^2 and 2000×1000 mm^2.

The optimization of the cutting process is vital for the company, since most of the waste stems from the cutting division. The scheme with the representation of the rectangular components of the vivaria in the stock sheet is called cutting pattern. The cutting patterns must be performed uninterruptedly across the stock sheet and parallel to one of its edges - this type of cut is called guillotine. Besides the guillotine constraint, the cutting patterns also have a limit on the number of stages (set of cuts with the same orientation - horizontal or vertical in which the stock material can be cut). The maximum number of stages allowed by the company is two plus an extra cut named trimming, which consists in separating only the rectangular components from waste material. An example of a cutting pattern of the type two-stage with trimming is presented in Fig.2.

The limitation on the number of stages of a cutting pattern is strongly related with the duration of the cutting process, since an increase in the number of stages,

Fig. 2 Two-stage cutting
pattern with trimming

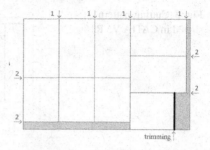

increases the complexity of the cutting pattern and the cutting time of the material. Although the supplier of cold rolled steel sheets can provide three different dimensions, the dimension most commonly used is $2500 \times 1250 \, \text{mm}^2$. The cutting patterns currently used by the company are composed by only one type of component. This type of cutting pattern is called homogeneous.

From the description of the cutting process adopted by the company, the amount of waste was evident as well as the need for over production in the adopted cutting patterns.

3 Solution Approach

The challenge in this case study was to develop a tool to optimize the cutting process, which could be easily used at the operational level by workers that are unaware of the operations research field, namely the problems definition and the resolution methods. The approach to this challenge was to use Microsoft Office Excel, since it was software that was already in the company and that users were familiar with.

The definition of the problem and the main objective for the optimization of the cutting process was developed in close collaboration with the company. However, it was not a simple task, since the main objective was not clear. On one hand, the company wanted cutting patterns that could be cut quickly, that is, which were not complex on the number of stages and, on the other hand, they intended to minimize the total waste and the number of plates used. In order to deal with these issues, three different objectives were considered: the minimization of the number of cuts, the minimization of the waste material and the minimization of the number of cold rolled sheets used.

Concerning the method, we adopted the mathematical formulation proposed by Gilmore & Gomory (1965) as follows. Let N be the set of component types required to produce a vivarium. Each component i, $i \in N = 1, \ldots n$ has a demand d_i, a length l_i and a width w_i. It is assumed that there are enough cold rolled steel sheets to produce all the required components. Each cold rolled steel sheet has length L and width W.

It is assumed that the set P of all cutting patterns is known. Associated to each cutting pattern p, there is a vector $(a_{1p}, a_{2p}, \ldots, a_{np})$ where a_{ip} is the number of times the component i is considered in the cutting pattern p. To each cutting pattern

Fig. 3 Cutting pattern designed in CATIA V5 R16

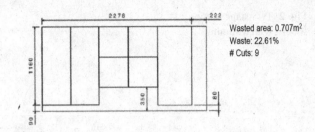

p is also assigned t_p, the number of cuts required to obtain all the components of the cutting pattern p, and s_p the waste of raw material associated with the cutting pattern p (that is, $s_p = LW - \sum_{i \in N} l_i w_i a_{ip}$). Let c_p be the coefficient of the objective function associated with the cutting pattern p, that is equal to t_p if the goal is to minimize the number of cuts, or to s_p if it intends to minimize the waste, or finally, equal to 1 if the objective is to minimize the number of cold rolled steel sheets used. The mathematical formulation for the problem is:

$$\text{Minimize} \quad \sum_{p \in P} c_p x_p \tag{1}$$

$$\text{Subject to:} \quad \sum_{p \in P} a_{ip} x_p \geq d_i, \qquad \forall i \in N; \tag{2}$$

$$x_p \geq 0 \text{ and integer}, \qquad \forall p \in P. \tag{3}$$

Each decision variable x_p corresponds to the number of times the cutting pattern p is cut. Usually the set P of all cutting patterns is large and the problem is difficult to be solved to optimality. In the problem considered in this case study, there was a major concern by the company in the validation of the cutting patterns. Since this point was very relevant, the strategy adopted was the use of the software CATIA V5 R16 for the representation of the cutting patterns, enabling the visual validation of the solutions obtained. An example of a cutting pattern developed in CATIA V5 R16 is presented in Fig. 3. For each cutting pattern p, the parameters to be used in each objective function, the number of cuts t_p and the total waste s_p, were calculated.

In summary, the solution approach is composed of two-phases, firstly the cutting patterns are designed in the software CATIA V5 R16 and validated by the company, secondly the vectors of each cutting pattern $(a_{1p}, a_{2p}, \ldots, a_{np})$, are added to Microsoft Office Excel and the objective function is selected, and the model (1)-(3) is solved.

4 Computational Results

The solution approach proposed was tested with the vivarium with the highest sales, as mentioned in section 2. In Table 1 the dimensions of the vivarium are presented. In the computational experiments the three different sizes of the cold rolled steel

Table 1 Dimension of the components

Component	Length (mm)	Width (mm)	#
Back	1170	518	1
Top	1160	555	1
Bottom	1160	430	1
Side	540	455	2
Division	450	450	1
Board	613	520	2

sheets ($3000 \times 1500\,\text{mm}^2$, $2500 \times 1250\,\text{mm}^2$ and $2000 \times 1000\,\text{mm}^2$) were tested. A demand of 80 vivaria was considered, since it is the amount in a typical order.

The dimension 2500×1250 mm^2 is the one most frequently used therefore the design of the cutting patterns begins with this sheet type. A total of 86 cutting patterns were created and validated by the company. When the dimension 3000×1500 mm^2 was analysed, 74 additional cutting patterns were produced and for the dimension 2000×1000 mm^2, 27 additional cutting patterns were added in Microsoft Office Excel. Thirteen tests were performed, for the three different objective functions,

$$f(x) = \sum_{p \in P} s_p x_p,$$

$$g(x) = \sum_{p \in P} t_p x_p,$$

$$h(x) = \sum_{p \in P} x_p,$$

that represent the minimization of waste $f(x)$, the number of cuts $g(x)$ and the total number of sheets $h(x)$, for the three different dimensions of sheets and for a version that considers the use of multiple sheet sizes simultaneously. Since it was not clear which of the objective functions was better suited to the problem, the three objectives were considered simultaneously, through different weights: $100f(x) + g(x) + 10h(x)$, with greater weight attributed to waste, followed by the number of sheets and the lowest weight to the number of cuts, considering the multiple sheet sizes.

In Fig. 4 the results are summarized, the total waste, the total number of cuts and the total number of sheets are compared, when different objective functions are considered. The current solution, being practised in the company, which considers the cold rolled sheets of dimension 2500×1250 mm^2 and only homogeneous patterns, is also represented for comparison.

The dimension 2000×1000 mm^2 is not a good option, since this dimension obtained the worst results in all the tests performed. It is likely that this sheet size does not have a good geometric combination with the dimensions of the components. On the other hand, using sheets of multiple sizes obtained the best results considering the different analysis for wastage, number of cuts and number of sheets.

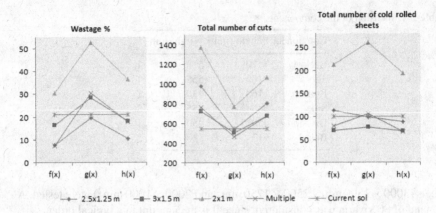

Fig. 4 Computational results for an order of 80 vivaria

Table 2 Computational results for the weighted objective function

Sheet Dimension	Objective function	Wastage %	Total number of cuts	Total number of cold rolled steel sheets
2500×1250 mm^2	Unweighted	7.6	536	88
	Weighted	10.4	689	88
Multiple	Unweighted	7.2	467	67
	Weighted	7.4	733	77
Current sol.		21.1	544	100

Since good results were obtained when multiple stock sheets were considered, and the dimension 2500×1250 mm^2 is the most commonly used in practice, the weighted objective function was used for these sizes. The results are summarized in Table 2; the weighted objective function is compared with the best results obtained for $f(x)$, $g(x)$ and $h(x)$, and with the current solution in practice at the company. In the weighted objective function, the wastage and the total number of sheets have higher impact, which is reflected in the results. The solutions obtained with the weighted objective function are competitive with the unweighted version, except for the total number of cuts, in which the solutions obtained were much worse.

Annual orders for the vivarium considered are around 1500 units, therefore computational experiments were performed, using these values as demand. The dimension 2500×1250 mm^2 was used, since it is the dimension presently in use at the company allowing a direct comparison with the current solution. Additionally, in the first computational experiments it was concluded that using multiple cold rolled sheet sizes enables good results in the three objective functions and thus multiple sheets are also considered for the instance with demand 1500.

The results are summarized in Fig.5. Analysing the different objectives independently, the use of multiple sheets is the best option with the smallest value for waste,

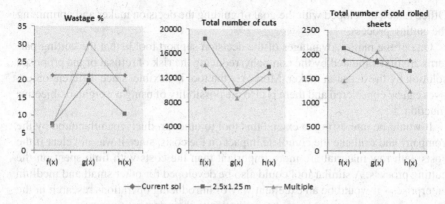

Fig. 5 Computational results considering the annual demand of the vivarium

Table 3 Quality of the solutions

| Problems | | | | | Gilmore & |
Sheet Dimension (mm²)	Demand of vivaria	Company solution	Proposed algorithm	Silva et al. (2010)	Gomory (1965)
2500×1250	80	100	88	88	88
2000×1000	80	-	194	160	160
3000×1500	80	-	68	61	60.33
2500×1250	1500	1875	1650	1650	1650

when the objective is to minimize waste $f(x)$, with the smallest number of cuts when the objective function is $g(x)$ and obtaining the smallest amount of sheets when the objective function is $h(x)$.

In order to analyse the quality of the solutions obtained by the proposed approach, the solutions obtained for the objective function $h(x)$ are compared with the optimal solutions obtained by the integer programming model proposed by Silva et al. (2010) and with the linear relaxation obtained by applying the column generation technique of Gilmore & Gomory (1961) for the cutting stock problem. The final results are presented in Table 3. Comparing with the optimal solutions, good results were obtained, with the exception of the stock sheet with dimension $2000 \times 1000 \, \text{mm}^2$ and demand 80 for the vivarium components.

5 Conclusions

In this paper we analysed the cutting process in a small Portuguese company. An analysis of the performance of the cutting process of one product was conducted, considering different sizes of the raw material and different objective functions. The output of this case study was a decision support tool implemented in Microsoft

Office Excel, developed with the goal of guiding the decision maker and optimizing the cutting process.

One of the major advantages of the decision support tool is that the cutting patterns are first validated by the company, reducing the risk of refusal of the proposed solution by the decision maker. Moreover the tool is flexible, since different objectives can be considered and there is also the possibility of using a weighted objective function.

It would be interesting to extend this tool to other products manufactured by the company and evaluate the complete impact on the costs, since it was not clear if the costs with raw material are more important than the costs with time spent on the cutting process. A similar tool could also be developed for other small and medium enterprises - it would be an excellent way of introducing operations research in the companies and give an important contribution to their competitiveness.

Acknowledgements. The second author is grateful to FCT — Fundação para a Ciência (Portuguese Foundation for Science and Technology) - for awarding the grant SFRH/ BPD/98981/2013.

References

Gilmore, P.C., Gomory, R.E.: A linear programming approach to the cutting-stock problem. Operations Research 9(6), 849–859 (1961)

Gilmore, P.C., Gomory, R.E.: Multistage cutting stock problems of two and more dimensions. Operations Research 13(1), 94–120 (1965)

Silva, E., Alvelos, F., de Carvalho, J.M.V.: An integer programming model for two-and three-stage two-dimensional cutting stock problems. European Journal of Operational Research 205(3), 699–708 (2010)

Wäscher, G., Haußner, H., Schumann, H.: An improved typology of cutting and packing problems. European Journal of Operational Research 183(3), 1109–1130 (2007)

Mathematics/Operations Research and Temperature on Cancer: A First Overview

Ana Paula Teixeira[1,2] and Regina de Almeida[1,3]

[1] Department of Mathematics, University of Trás-os-Montes e Alto Douro (UTAD),
5000 Vila Real, Portugal
[2] CIO–Operations Research Center, FCUL - University of Lisbon, 1749 Lisbon, Portugal
ateixeir@utad.pt
[3] CMAT, University of Minho, Campus de Gualtar, 4710-057 Braga, Portugal
ralmeida@utad.pt

Abstract. In this study, the main characteristics of research works on Mathematical /
Operations Research techniques applied to cancer and involving the temperature are
described and analyzed. The information contained in those works was cataloged in
accordance with specific key elements previously chosen. The developments on the
theme are evaluated, observing what was done, describing the related applications,
procedures and their practical implementation.

Keywords: Cancer modeling, Operations Research techniques.

1   Introduction

Some of the earliest mathematical contributions to the study of tumors date from
the late 20's and 30's, such as (Hill 1928) study, although not specifically applied
to the study of cancer, leaded for many later mathematical models of solid tumors
and the work of (Mayneord 1932) on the effects of X-radiations on sarcoma.
Consequence of the radiation physics and computer technology advances, several
works relating to radio sensitivity of tumors appeared in the 80´s and 90's, such as
the ones of (Schultz and King 1987) and (Levine *et al.* 1995), among others. The
21st century witnessed an explosion in the publication of mathematical papers that
apply techniques based on the temperature on tumor analysis, such as (Yang *et al.*
2000), (Maksimova *et al.* 2007), (Gnyawali *et al.* 2008), (Nemkov *et al.* 2011) and
(Aghayan *et al.* 2013).

Although it is possible to find in the literature valuable and interesting synthesis
and reviews on the application of mathematical techniques applied to cancer
analysis, like the ones by (Byrne 2010) and (Bellomo *et al.* 2008), we consider
that further analysis on this subject is still needed, in order to better understand
what has been done in this area, specifically aiming to identify which Operations
Research (OR) techniques are used in the study of different types of tumors, to

© Springer International Publishing Switzerland 2015

A.P.F.D. Barbosa Póvoa and J.L. de Miranda (eds.), *Operations Research and Big Data*,
Studies in Big Data 15, DOI: 10.1007/978-3-319-24154-8_27

gain insight on the future directions of this field and draw some improvements from the most recent developments. Here, we outline a selection of studies to illustrate how this field of research has taken shape and how OR techniques have contributed to an enhanced understanding of tumor development over the recent years. In Section 2, the methodology is introduced; in Section 3, the results of this overview are described; and in Section 4, the conclusions are presented.

2   Methodology of the Study

Researching the "Mathematical techniques applied to cancer" we can encounter a wide field of study. Thus, we must delimit it so that the quantity and diversity of data do not prevent us from achieving our purposes. Additionally, we are only interested in those works that can be considered essential on Mathematical and Health fields. Therefore, a set of relevant databases and web search engines were selected, namely Web of Science, Current Contents Connect, Inspec, PubMed, Academic Research Microsoft and Scholar Google, since they contain a wide range of top journals. Furthermore, our attention goes essentially to specialized works published on books, journals, proceedings, as well as MSc and Ph.D. thesis and working papers, completing the selection with other relevant publications in this area. In this article we focuses on specialized works, written in English, in Mathematical/OR techniques applied to cancer analysis involving the temperature that were obtained from the Scholar Google engine.

The analyzed works were cataloged in accordance with seven specific key elements, chosen taking into account both the topics that have been explored in the literature over almost a century and the methodological approach suggested by many authors of review papers: 1) Scientific areas involved: Health Sciences (HS), Sciences and Technology (ST). 2) Activities performed: like characterize, analyze, determine. 3) Resources / Techniques involved: imagiologic evaluation (like tomography, mammography, X-ray, radiology), treatment approaches (like cryosurgery, hyperthermia, laser, radiofrequency), diagnosis techniques (like biopsy, histology, endoscopy, thermography) and resources (like, microscopic, spectrometer, infrared camera). 4) Type of mathematical model: computational, continuous, discrete, hybrid, agent based, dynamic, deterministic and stochastic. 5) Mathematical methodology applied: OR (like optimization, simulation and heuristics) and non-OR (for example, statistics, numerical analysis). 6) Type of tumor: like brain, breast, colorectal, generic. 7) Type of work: review, theoretical, practical implementation with generated and/or theoretical data and practical implementation with real data. Moreover, the developments on the application of Mathematical / OR techniques to cancer involving the concept of temperature are evaluated, observing what has been done, describing the related applications and procedures, and exploring their practical implementation.

A document search using the terms "Temperature", "Cancer" and "Mathematics" in the title, abstract or keywords of the analyzed works, as well as a manual search to detect synonyms of the above expressions, was performed.

In the next phase, the references cited in the works obtained, as well as other items listed in the search engines, were also analyzed. After this process, both authors individually examined each article and identified the most important elements; if there was a discrepancy in the categorization of articles reviewed a reassessment was performed until there was a consensus.

3 Results of the Study

Some reviews on mathematical applications to cancer have already been published. Here, we briefly mention two of those works: (Byrne 2010) published a review of the literature on progress in mathematical modeling of cancer since the 60's; the different theoretical approaches that have been applied to tumors and the novelties and improvements that have arisen are focused; in particular, mathematical models of carcinogenesis, avascular and vascular tumor growth and angiogenesis; and (Bellomo *et al* 2008) present a critical review of some topics related to mathematical approaches to the modeling of phenomena in cancer dynamics, with the aim of illustrating, to applied mathematicians, some of the recent mathematical problems in the field.

Regarding the Scholar Google engine, about thirty seven works on Mathematical / OR techniques applied to cancer and involving the temperature were obtained after the analyzis process; two of them are books and three are reviews (two of the reviews are shortly described above). Among the other thirty two works, our focus goes to the eighteen ones which use OR techniques, wherein almost 90% of them are articles published in scientific journals, being the remaining 10% conference proceedings and working papers. Considering the space limitations, we decided to present here a synthesis describing only some of the contributions of the analyzed works, mainly in terms of the most used OR techniques, aiming to present a more detailed review on this topic in a near future.

Table 1 The two most cited categories of each key element

Key elements	Percentages of the categories			
Scientific areas	HS and ST	89%	Just ST	11%
Activities	Characterize	50%	Analyze	28%
Health Techniques	Each: Treatment approaches / Resources	33%	Each: Imagiologic evaluation / Diagnosis techniques	17%
Type of Model	Computational	100%	Continuum	50%
Mathematic Techniques	Simulation	72%	Optimization	33%
Type of Works	Real data	56%	Non real data	44%
Type of Tumors	Generic	33%	Each specific type *	11%

* *pelvic, breast, prostate, skin* and *oral mucosa*

All the analyzed works were classified in accordance with the key elements described in Section 2. In Table 1 we present the two most frequently cited categories related to each one of those elements, for the eighteen works under study. Of those, 33% focus on tumors in general ((Ramis-Conde *et al.* 2008),

(Turner *et al.* 2004), (Lang *et al.* 1999), (Aghayan *et al.* 2013), (Gerlee and Anderson 2008), (Bagheri *et al.* 2011)), while *pelvic, breast, prostate, skin* and *oral mucosa* cancers are explored by 11% of the works each ((Van den Berg *et al.* 1983), (Deuflhard and Seebass 1998)), ((Pennisi *et al.* 2008), (Gnyawali *et al.* 2008)), ((Oden *et al.* 2006), (Nemkov *et al.* 2011)), ((Maksimova *et al.* 2007), (Gnyawali *et al.* 2008)) and ((Maksimova *et al.* 2007), (Wang and Garibaldi 2005)), respectively). Almost all the works involve both *Health Sciences* and *Sciences and Technology* areas, being (Gerlee and Anderson 2008), (Wang and Garibaldi 2005) the only ones involving just the area of *Sciences and Technology*. *Characterize* and *Analyze* are the most frequent *activities* ((Deuflhard and Seebass 1998), (Oden *et al.* 2006), (Maksimova *et al.* 2007), (Nemkov *et al.* 2011), (Ramis-Conde *et al.* 2008), (Yang *et al.* 2000), (Lang *et al.* 1999), (Zhang *et al.* 2009b), (Wang and Garibaldi 2005)) and ((Van den Berg *et al.* 1983), (Turner *et al.* 2004), (Basdevant *et al.* 2005), (Zhang *et al.* 2009a), (Gerlee and Anderson 2008)), respectively), while *Treatment approaches, Imagiologic evaluation* and *Diagnosis techniques* are the most used *techniques* (((Van den Berg *et al.* 1983), (Deuflhard and Seebass 1998), (Gnyawali *et al.* 2008), (Yang *et al.* 2000), (Lang *et al.* 1999), (Aghayan *et al.* 2013)), ((Van den Berg *et al.* 1983), (Oden *et al.* 2006), (Nemkov *et al.* 2011)) and ((Maksimova *et al.* 2007), (Gnyawali *et al.* 2008), (Turner *et al.* 2004)), respectively). Additionally, *Resources* are specified by 33% of the works ((Gnyawali *et al.* 2008), (Turner *et al.* 2004), (Basdevant *et al.* 2005), (Yang *et al.* 2000), (Lang *et al.* 1999), (Wang and Garibaldi 2005)). The OR techniques quoted are mainly under the scope of *Simulation* and *Optimization*. Concerning the references involving *Simulation*, (Van den Berg *et al.* 1983), (Deuflhard and Seebass 1998), (Oden *et al.* 2006), (Maksimova *et al.* 2007), (Gnyawali *et al.* 2008) and (Bagheri *et al.* 2011) use *deterministic dynamic continuum* techniques; (Zhang *et al.* 2009a, 2009b) use *deterministic dynamic agent based* techniques, (Gerlee and Anderson 2008) use *deterministic dynamic hybrid* techniques; while (Ramis-Conde *et al.* 2008) and (Turner *et al.* 2004) use *stochastic dynamic hybrid* techniques and *stochastic dynamic discrete* techniques, respectively; (Maksimova *et al.* 2007) also uses *stochastic dynamic continuum* techniques. Within the papers where *Optimization* techniques are used, (Oden *et al.* 2006), (Basdevant *et al.* 2005), (Lang *et al.* 1999) and (Aghayan *et al.* 2013) focus on *Non Linear Programming*, whereas (Pennisi *et al.* 2008) and (Wang and Garibaldi 2005) use *Global Optimization* as well as *local-search Meta-heuristics*. All the models in the reviewed literature are *Computational*, being half of them *Continuum* ((Van den Berg *et al.*1983), (Deuflhard and Seebass 1998), (Oden *et al.* 2006), (Maksimova *et al.* 2007), (Gnyawali *et al.* 2008), (Basdevant *et al.* 2005), (Lang *et al.* 1999), (Aghayan *et al.* 2013) and (Bagheri *et al.* 2011)). Slightly more than half of the works performed tests with real data ((Oden *et al.* 2006), (Maksimova *et al.* 2007), (Nemkov *et al.* 2011), (Gnyawali *et al.* 2008), (Basdevant *et al.* 2005), (Yang *et al.* 2000), (Lang *et al.* 1999), (Zhang *et al.* 2009b), (Wang and Garibaldi 2005), (Bagheri *et al.* 2011)).

A brief synthesis of some of the analyzed works on OR techniques applied to cancer and involving the temperature, published in prestigious journals, will follow. We will begin by the works using simulation techniques and then consider the ones regarding optimization (two of this last group also use heuristics).

In (Van den Berg et al.1983), computer simulation is used to analyze the potentialities of hyperthermia as a cancer treatment approach. First, the domain-integral-equation technique is used to compute the heat generated in a distribution of biological tissue when irradiated by a source of electromagnetic radiation; then, an iterative minimization of the integrated square error is used to solve numerically the obtained integral equation. Using a finite-element technique in space and a finite-difference technique in time to solve numerically the pertaining heat transfer problem, the temperature distribution follows from the computed distribution of generated heat. A mathematical model that describes the cancer therapy hyperthermia, involving Maxwell's equations in inhomogeneous media and a parabolic bioheat transfer equation, is presented in (Deuflhard and Seebass 1998), an adaptive multilevel finite element model for the Maxwell's equation, which dominates the numerical simulation time is used. Applications of silica(core)/gold(shell) nanoparticles to photothermal therapy of spontaneous tumor of cats and dogs are described in (Maksimova et al. 2007). Simulation of the spatial distribution of light absorbance related to the process of electromagnetic wave propagation through a system of discrete scattering particles with consideration of multiple scattering effects was carried out. To calculate the spatial distribution of the temperature, a two-dimensional Poisson equation was employed as a mathematical model. In (Nemkov et al. 2011), an inductor design optimization, using alternating magnetic fields, is presented. Simulation and in vitro test experiments are described. To attempt acquiring temperature distribution, in order to guide laser photo thermal treatment of tumors, in (Gnyawali et al. 2008), infrared thermography was used, to measure surface temperature during laser irradiation, and studies of the surface temperature using Monte Carlo simulation were performed. In (Ramis-Conde et al. 2008), a mathematical multiscale model simulation is used to show how cell adhesion may be regulated by interactions between E-cadherin and -catenin and how the control of cell adhesion may be related to cell migration, to the epithelial-mesenchymal transition and to invasion in populations of eukaryotic cells. An extended Potts model is used, in (Turner et al. 2004), to investigate the effects of TGF on tumor morphology and invasiveness, and also the implications that this growth hormone has for .tamoxifen treatment. Simulation results are presented. In (Yang et al. 2000), a new in vitro experimental model that records temperature changes over a culture plate and that can be used to assess the biological effects of cryosurgery is described. Temperature changes during freezing and thawing in a culture plate were monitored using a computer control system, and the data were used to create a temperature profile of the entire plate. Temperature changes at any point in the interpolated areas were estimated using a curve fitting method. The estimated temperature was checked by sampling with four additional randomly placed thermocouples. Linear regression analysis showed that the estimated temperature and measured temperature were very close. A multi-scale, multi-resolution, agent based, computer simulation glioma model is presented in (Zhang et al. 2009a), allowing to conclude that by pursuing a multi-resolution approach, the computation time of a discrete-based model can be substantially reduced while still maintaining a comparably high predictive power. In (Gerlee and Anderson 2008), simulation experiments with a cellular automaton model of clonal evolution in cancer are performed, with the aim of investigate the emergence of the glycolytic phenotype. In (Zhang *et al.* 2009b), an agent-based modeling is used in a multi-scale tumor modeling platform that conceives brain cancer as a complex dynamic biosystem. Simulation is performed and results are presented and discussed. Combinatorial treatment strategies using a mathematical model that predicts the impact of MEK inhibition on tumor cell proliferation, ONYX-015 infection, and oncolysis are studied in (Bagheri *et al.* 2011); a nonlinear differential equation system is fit to experimental data and the resulting simulations for favorable treatment strategies are analyzed. Simulations were successfully validated in an ensuing explicit test study and predicted enhanced combinatorial therapy when both treatments were applied simultaneously. A dynamic data-driven planning and control system for cancer laser treatment is developed in (Oden *et al.* 2006). Adaptive-feedback control of mathematical and computational models based on a posteriori estimates of errors in key quantities of interest and the use of magnetic resonance temperature imaging and diode laser devices to monitor treatment of tumors in laboratory animals are the essence of the proposed systems. Mathematically and numerically optimal strategies in cancer chronotherapy are studied in (Basdevant *et al.* 2005). Two different optimization problems are examined; the eradication problem, which consists in finding the drug infusion law able to minimize the number of tumor cells while preserving a minimal level for the villi population; and the containment problem, that searches for a quasi-periodic treatment able to maintain the tumor population at the lowest possible level, while preserving the villi cells. An optimal control technique is applied to search for the best drug infusion laws. In order to predict the temperature, i.e. to achieve desired steady–state temperature distributions, in (Lang *et al.* 1999) an optimization process based on a three–dimensional nonlinear heat transfer model, designed for regional hyperthermia of deep seated tumors, is described. In (Aghayan *et al.* 2013) an optimum tissue heating condition during hyperthermia treatment is proposed. A methodology to numerically optimize hyperthermia treatments based on an inverse heat conduction problem, by using the conjugate gradient method with an adjoint equation, is developed. The finite difference time domain method is applied to numerically solve the tissue temperature distribution using Pennes bioheat transfer equation. Simulated annealing was used, in (Pennisi *et al.* 2008), to downsize the computational effort, in order to find optimal solutions for drug or vaccine protocols that use system biology modeling. Biologically driven heuristic strategies are used to fasten the algorithm. To improve the clustering technique, in (Wang and Garibaldi 2005) a new Simulated Annealing Fuzzy Clustering method is proposed and its performance is evaluated on seven oral cancer tissue samples obtained through Fourier Transform Infrared Spectroscopy.

4   Conclusions

Throughout this research work, we observed a significant number of studies on cancer approaches using the temperature that involve both the Health Sciences and the Sciences and Technology areas, being that most of them apply simulation techniques with either real or generated data. Optimization methodologies are also used in some studies, being computational continuum models the most frequently found in the literature. The characterization of cancer treatment approaches, diagnosis techniques, imagiologic evaluations and used resources are one of the main concerns of the analyzed works and while some authors focus on approaches of tumors in general, others study a specific type of cancer.

Additionally, we realized that a mix of new strategies, experimental techniques and theoretical approaches are continuously emerging in the ongoing battle against cancer; thus, novel combinations of ingenious experimental designs and intelligent mathematical models and techniques will be imperative to clarify these and other enigmas; allowing us to conclude that there is still much work to be done in this area. We this in mind, we continue to deepen the study of this issue in order to present a broader overview in a near future.

Acknowledgments. The authors thank Dr.ᵃ Marcília Teixeira Mateus (Júlio Dinis Maternity) for the medical expertise advises and help concerning the health science area and also to UTAD University.

References

Aghayan, S.A., Sardari, D., Mahdavi, S.R.M., Zahmatkesh, M.H.: An inverse problem of temperature optimization in hyperthermia by controlling the overall heat transfer coefficient. Journal of Applied Mathematics, 1–9 (2013)

Bagheri, N., Shiina, M., Lauffenburger, D.A., Korn, W.M.: A dynamical systems model for combinatorial cancer therapy enhances oncolytic adenovirus efficacy by MEK-inhibition. PLoS Computational Biology 7(2), e1001085 (2011)

Basdevant, C., Clairambault, J., Lévi, F.: Optimisation of time-scheduled regimen for anti-cancer drug infusion. ESAIM: Mathematical Modelling and Numerical Analysis 39(06), 1069–1086 (2005)

Bellomo, N., Li, N.K., Maini, P.K.: On the foundations of cancer modelling: selected topics, speculations, and perspectives. Mathematical Models and Methods in Applied Sciences 18(04), 593–646 (2008)

Byrne, H.M.: Dissecting cancer through mathematics: from the cell to the animal model. Nature Reviews Cancer 10(3), 221–230 (2010)

Deuflhard, P., Seebass, M.: Adaptive multilevel FEM as decisive tools in the clinical cancer therapy hyperthermia. ZIB (1998)

Gerlee, P., Anderson.: A hybrid cellular automaton model of clonal evolution in cancer the emergence of the glycolytic phenotype. J. Theor. Biol. 250(4), 705–722 (2008)

Gnyawali, S.C., Chen, Y., Wu, F., Bartels, K.E., Wicksted, J.P., Liu, H., Chen, W.R.: Temperature measurement on tissue surface during laser irradiation. Medical & Biological Engineering & Computing 46(2), 159–168 (2008)

Hill, A.V.: The diffusion of oxygen and lactic acid through tissues. R Soc. Proc. B 104, 39–96 (1928)

Lang, J., Erdmann, B., Seebass, M.: Impact of nonlinear heat transfer on temperature control in regional hyperthermia. IEEE Transactions on Biomedical Engineering 46(9), 1129–1138 (1999)

Levine, E.L., Renehan, A., Gossiel, R., Davidson, S.E., Roberts, S.A., Chadwick, C., Wilks, D.P., Potten, C.S., Hendry, J.H., Hunter, R.D.: Apoptosis, intrinsic radiosensitivity and prediction of radiotherapy response in cervical carcinoma. Radiother. Oncol. 37, 1–9 (1995)

Maksimova, I.L., Akchurin, G.G., Khlebtsov, B.N., Terentyuk, G.S., Akchurin, G.G., Ermolaev, I.A., Tuchin, V.V.: Near-infrared laser photothermal therapy of cancer by using gold nanoparticles: Computer simulations and experiment. Medical Laser Application 22(3), 199–206 (2007)

Mayneord, W.V.: On a law of growth of Jensen's rat sarcoma. Am. J. Cancer 16, 841–846 (1932)

Nemkov, V., Ruffini, R., Goldstein, R., Jackowski, J., DeWeese, T.L., Ivkov, R.: Magnetic field generating inductor for cancer hyperthermia research. COMPEL-The International Journal for Computation and Mathematics in Electrical and Electronic Engineering 30(5), 1626–1636 (2011)

Oden, J.T., et al.: Development of a computational paradigm for laser treatment of cancer. In: Alexandrov, V.N., van Albada, G.D., Sloot, P.M.A., Dongarra, J. (eds.) ICCS 2006. LNCS, vol. 3993, pp. 530–537. Springer, Heidelberg (2006)

Pennisi, M., Catanuto, R., Pappalardo, F., Motta, S.: Optimal vaccination schedules using simulated annealing. Bioinformatics 24(15), 1740–1742 (2008)

Ramis-Conde, I., Drasdo, D., Anderson, A.R., Chaplain, M.A.: Modeling the influence of the E-cadherin-β-catenin pathway in cancer cell invasion: a multiscale approach. Biophysical Journal 95(1), 155–165 (2008)

Schultz, D.S., King, W.E.: On the analysis of oxygen diffusion in biological systems. Math. Biosci. 83, 179–190 (1987)

Turner, S., Sherratt, J.A., Cameron, D.: Tamoxifen treatment failure in cancer and the nonlinear dynamics of TGFβ. J. Theor. Biol. 229(1), 101–111 (2004)

Van den Berg, P., De Hoop, A., Segal, A., Praagman, N.: A computational model of the electromagnetic heating of biological tissue with application to hyperthermic cancer therapy. IEEE Transactions on Biomedical Engineering, (12) 797–805 (1983)

Yang, W., Peng, H., Chang, H., Shen, S., Wu, C., Chang, C.: An in vitro monitoring system for simulated thermal process in cryosurgery. Cryobiology 40(2), 159–170 (2000)

Wang, X.Y., Garibaldi, J.M.: Simulated Annealing Fuzzy Clustering in Cancer Diagnosis. Informatica (Slovenia) 29(1), 61–70 (2005)

Zhang, L., Chen, L.L., Deisboeck, T.S.: Multi-scale, multi-resolution brain cancer modeling. Mathematics and Computers in Simulation 79(7), 2021–2035 (2009)

Zhang, L., Wang, Z., Sagotsky, J.A., Deisboeck, T.S.: Multiscale agent-based cancer modeling. Journal of Mathematical Biology 58(4-5), 545–559 (2009)

Performance Assessment of Children and Youth Households

Clara Bento Vaz [1], Jorge Alves[2], and Ivo Mendes[3]

[1] Instituto Politécnico de Bragança, CGEI/INESC TEC, Portugal
[2] Instituto Politécnico de Bragança, OBEGEF/UNIAG, Portugal
[3] Instituto de Segurança Social, I.P., Coimbra, Portugal

Abstract. This study purposes a Data Envelopment Analysis (DEA) framework to assess the performance of Non-Profit Organizations that look after children and young people. The DEA method is used to assess the managerial efficiency of eight institutions from Bragança district, since 2010 to 2013. The model evaluates each institution concerning the reduction of operational and staff costs incurred in providing social services to the number of users observed in each unit. A fair policy for the allocation of subsidies is designed according to the performance of institutions in order to support the regulator.

Keywords: Third Sector, Children and Youth Households, Efficiency, DEA.

1 Introduction

The Third Sector (TS) in Portugal has gained increasing strength, being in charge of traditional social services of the State sphere. The transference of these traditional services from the public to the TS implied the allocation of substantial economic support which should be managed with extreme rigour. Therefore, it is particularly important that public structures, which support the Non-Profit Organizations (NPO), have knowledge of how efficiently and effectively those NPO manage the available funds and the resources [1]. As argued by Vakkuri [2], Sinuany-Stern and Sherman [3], DEA is a good methodology for measuring performance in NPO, although there are few studies in TS [4]. The intrinsic nature of DEA in determination an efficiency measure taking into account that each organization uses multiples resources to produce multiple outputs has increased the number of applications in TS, hindering the use of traditional ratios. These ratios measure partial efficiency as for example the ratio of total subsidy to total amount spent on social service [5]. Other approaches determine the overall efficiency taking a weighted aggregation of individual variables, using predefined weights subjectively determined for combining variables.

The study purposes a DEA framework to support the assessment of managerial efficiency of NPO looking after children and young people that are referred to

© Springer International Publishing Switzerland 2015

A.P.F.D. Barbosa Póvoa and J.L. de Miranda (eds.), *Operations Research and Big Data,*
Studies in Big Data 15, DOI: 10.1007/978-3-319-24154-8_28

237

hereon as Children and Youth Households (CYH). This approach intends to elucidate the regulator authority and the community in management and supervising
the NPO in terms of efficiency in using the subsidies received and the resources
required to provide the social services which is an example of community-based
operations research as referred by Sinuany-Stern and Sherman [3]. The first contribution of this paper is to assess the managerial efficiency of each CYH in reducing the cost of resources required to take care of the number of the available users.
The second contribution is to propose a fair policy for allocating the subsidies
based on DEA results.

The CYH is one of the social responses offered by the NPO in Portugal. These
institutions look after children and young people, being in critical situation for
more than 6 months, aged between 6 and 18 years. For children who are below 6
and for the ones with specific health problems there are other kind of institutions
that can receive them. Generally, the distribution of children by age is quite homogeneous in the various CYH observed. The objectives of this social response
are [6]: ensuring the accommodation and the needs of the children and young
people, promoting their overall development, in conditions as close as possible to
a family structure; ensuring the necessary resources for their personal development and vocational training, in cooperation with the family, school, professional
training and community; and promoting their integration in the family collaborat-
ing with the relevant authorities, to achieve the gradual children's autonomy from
the institutions.

This study presents a performance assessment of 8 CYH, from Bragança district, from 2010 to 2013, which corresponds to the available data. As far as we
know, there is no published study documenting DEA approach for assessment of
these social welfare institutions.

The paper is organized as follows: Section 2 reviews the literature about the
performance assessment methods for NPO in the TS. Section 3 describes the case
study and presents the DEA methodology. Section 4 describes the DEA model and
discusses the results. Section 5 concludes the paper.

2 Literature Review in Measuring Performance in Third Sector

Performance evaluation studies in the TS institutions are still scarce [1]. According to Sillanpää [7], the performance of these institutions can be evaluated from
the results of their activity such as financial results or also from the process of
how the activity is performed in terms of efficiency, quality and effectiveness.

The performance of TS organizations has been assessed through qualitative and
quantitative approaches [8]. However, in certain institutions or contexts more
importance is given to the qualitative performance rather than to the quantitative
performance obtained through the financial information [9]. Often it is also used a
combination of the two approaches (quantitative and qualitative) by using both
financial and non-financial criteria as exemplified by Kaplan and Norton's Balanced Score Card [10], the European Foundation for Quality Management

(EFQM) [11] and DEA [12]. In TS, the DEA applications are found in performance assessment of social service agencies [12], provision of social care for older people [13], education and human services public assistance [14] and international aid organizations [15].

This study presents a performance assessment of CYH looking after children and young people that can be used by the regulator in managing and allocating public funds for these institutions.

3 Performance Assessment of CYH

This section describes the case study and introduces the DEA methodology.

3.1 Case Study

The CYH are non-profit and not public organizations, which have social purposes and private management. The activity of CYH is supported by the State according to the number of users (children and young people) and the Cooperation Agreement (CA) established with the Social Security Institute (SSI). The CA is defined according to the technical capacity and physical conditions of each CYH and includes information about their duties, duration of agreement, user vacancies and the financial contribution per user. The subsidy received per user is the same in different CYH, for a given year. The subsidy received per user and month by each CYH was 503€ in 2010, it decreased to 484€ and 470€ in 2011 and 2012, respectively, and it increased to the interval ranging from 500€ to 700€, in 2013. In this year, each CYH received a total subsidy which corresponds to the total number of user vacancies contracted in CA if its occupation rate is at least 65%. This limit was 75% for 2009 and 2010 and it was reduced to 50% in 2011 and 2012. For a lower occupancy rate, the CYH only receives funds according to the number of users that they effectively accommodate. Each CYH should manage the received subsidies to cover the costs incurred in its activity. The amount of subsidies received by the institutions are usually higher than their costs. In the period studied, there are one or two institutions that, in some years, received subsidies which were lower than the total costs. This situation occurred in the CYH 7 that closed in 2013.

In 2007, SSI has implemented a quality management framework for the CYH based on the EFQM. This approach is used to verify if the social services have been correctly provided by each CYH, establishing a minimum level of requirements which have to be fulfilled [16]. Unfortunately, unavailability of results from the quality assessment of the CYH prevented the authors from complementing the proposed DEA approach. This quality management perspective only focuses on the social service provided to the user, disregarding the efficiency assessment of the CYH in resources and subsidies management. In this context, the evaluation of the managerial efficiency of each CYH is critical which enables to assess how each CYH manages the resources and the subsidies received in providing social services to the level of users in each institution. Thus, this paper proposes a DEA framework to evaluate the performance of CYH which should complement the current quality assessment model.

This study considers the panel dataset of 8 CYH from Bragança district, located in different municipalities, from 2010 to 2013. It was observed an outlier concerning the CYH 3, in 2012. In 2013, the CYH 7 is excluded from the assessment because it was closed. Thus, the final sample includes 30 observations, from the 8 CYH, excluding two observations regarding CYH 3 and 7. The mean and the coefficient of variation (c.v.) of the variables across CYH, in the corresponding year, are summarized in Table 1, considering that the monetary variables were deflated using devaluations coefficients with reference to 2013.

Table 1 Mean and the coefficient of variation values for the inputs, output and subsidies of CYH

Year	Inputs				Output		Subsidies (€)	
	Staff cost (€)		Operating costs (€)		Users			
	Mean	c.v.	Mean	c.v.	Mean	c.v.	Mean	c.v.
2010	189 387	37%	115 914	56%	28.5	52%	295 927	48%
2011	175 522	39%	97 297	55%	31.4	51%	292 872	47%
2012	186 435	33%	101 159	45%	32.6	44%	298 893	36%
2013	174 643	46%	105 799	80%	27.7	55%	294 792	58%

The higher scores of variability relative to the mean are observed for operating costs, followed by the number of users and subsidies. It is observed that the average total cost per user is higher in 2010 and 2013 than in other years because the average subsidies per user is also higher in those years. Actually, the institutions spend more money in their activities if they receive more subsidies. The DEA methodology is introduced in the next section.

3.2 DEA Methodology

The managerial efficiency for each CYH is evaluated through the technical efficiency derived from the DEA model, introduced by [17]. DEA is a non-parametric approach to assess the relative efficiency of a homogeneous set of Decision Making Units (DMUs) in producing multiple outputs from multiple inputs. DEA identifies a subset of efficient CYH considered as benchmarks. These DMUs define the frontier technology enveloping all institutions observed in the Production Possibility Set (PPS). For the inefficient CYH located inside the PPS, the magnitude of the inefficiency is derived by the distance to the frontier and a single summary measure of efficiency is calculated.

Consider a set of n CYH j $(j = 1, ..., n)$, each consuming m resources (inputs) x_{ij} $(x_{1j}, ..., x_{mj})$ to produce s results (outputs) y_{rj} $(y_{1j}, ..., y_{sj})$. For an input minimizing perspective and assuming the most productive frontier observed, defined by constant returns to scale (CRS), the relative efficiency of the assessed CYH_o can be evaluated using the linear programming model (1):

$$Min\{\theta_o | \theta_o x_{io} \geq \sum_{j=1}^{n} \lambda_j x_{ij}, i = 1, \ldots, m, \ y_{ro} \leq \sum_{j=1}^{n} \lambda_j y_{rj}, r = 1, \ldots, s, \lambda_j \geq 0\} \quad (1)$$

The relative efficiency (θ) of the assessed CYH_o is calculated by the optimum solution of model (1), θ_o^*, corresponding to the minimum factor by which the inputs levels can be reduced given the current level of outputs. DEA enables to identify the efficient CYH which have the best practices and the inefficient CYH which activity can be improved. The efficiency measure is equal to 100% when the CYH under assessment is efficient, whereas lower scores indicate the existence of inefficiencies. For inefficient CYH, it is also possible to obtain, as by-products of the DEA efficiency assessment, a set of targets for becoming efficient, which are feasible points observed on the frontier.

As the scale size affects the productivity of a CYH, it is important to calculate the scale efficiency to measure the distance between CRS and variable returns to scale (VRS) frontiers at the scale size of the assessed unit, through the ratio between θ_o^* and the efficiency score achieved assuming VRS, which requires including the constraint $\sum_{j=1}^{n} \lambda_j = 1$ in model (1). Thus, the larger the difference between CRS and VRS frontiers, the lower the value of scale efficiency is, and the adverse impact of scale size on productivity is more significative.

4   Results and Discussion

This section describes the DEA model application and the discussion of the results.

4.1   DEA Model

The managerial efficiency for each CYH is assessed by the model (1), evaluating its capacity in minimizing the operating and staff costs required to take care of the children and young people observed in the institution.

The single output of the model is the number of users in each CYH, calculated through the average monthly number of users, per year. The costs of each CYH essentially depend on the number of users. The children's age and the social care are quite homogeneous in the various CYH observed. The inputs of the model include the total cost of technical and education staff effectively spent by each CYH per year, and the total annual operating costs concerning external supplies and services (heating system, electricity, water and maintenance of institution), food and other consumed materials. These resources should be minimized for a given level of users accommodated in each CYH. The data concerning the inputs are collected from income statements regarding the 8 CYH which are provided by SSI and summarised in Table 1.

4.2 Empirical Results

The relative efficiency of a CYH in a given year is estimated by comparison to the best practices observed during the period analysed, ranging from 2010 to 2013. In order to see the impact of scale size on the efficiency in the period studied, we calculated the average of scale efficiency for each CYH which is higher than 81%, except for CYH 4 and 8 which is 51% and 76%, respectively. Thus, only these two institutions show inefficiency due to scale size which strengthens the use of CRS frontier technology to assess the CYH efficiency. According to the decision maker, the CRS assessment reflects in higher degree the performance expectations concerning each CYH.

The average managerial efficiency for the CYH is 64% (with a standard deviation of 19%), indicating that each CYH can reduce their current operating and staff costs by 36%, on average, providing the social services to the same level of users in each institution. This analysis revealed that there is only 2 efficient institutions, CYH 3 in 2011 and CYH 6 in 2010, being the benchmarks. The best practices observed in these institutions should be identified to be emulated by the inefficiency institutions with efficiency scores ranging from 29% to 93%. The results indicate that there are three inefficient CYH that have slack in the constraint relative to the input operating costs. This means that these inefficient CYH should make an extra effort in reducing the level of operating costs without detriment to their social services provided to the users. Looking now at the distribution of efficiency scores across institutions, given in Fig. 1, it is observed that in general the institutions have decreased the managerial efficiency over the last years. In fact, the average managerial efficiency of the CYH increased by 24% in 2011, decreasing by 6% in 2012 and 2013.

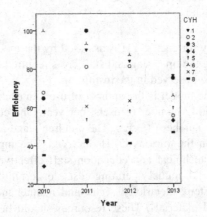

Fig. 1 Managerial Efficiency for each CYH per year.

A fair policy of allocating the subsidies is proposed by defining that the total subsidy received by each CYH should be proportional to the real number of users independently from the number of vacancies available in each institution. The allocating of the users and consequently the subsidies should be beneficiating the institutions which are more efficient in management the staff and operating costs given the available users based on the results from DEA model.

Taking into account the staff and operating costs observed on the two benchmarks, we propose a score for the subsidy per user and per month (S), which is calculated according to: $S = C_s(1 + P_s) + C_o(1 + P_o)$. C_s is the staff cost per user defined as the weighted average of the staff costs per user observed in the benchmarks. The same procedure is used to calculate the operating costs per user (C_o). The weight for each benchmark is the percentage of times used as benchmark, in the evaluation. P_s is the estimated percentage that each institution needs to be prepared, in terms of extra staff, for the possibility to receive more users. In the period studied, P_s is calculated by the ratio between the total number of vacancies available in all institutions and the total number of vacancies in all institutions defined according to the CA. P_o is the estimated percentage to account the possibility to have extra costs due to equipment depreciation and other costs, which is approximately 5% of the operating costs. In the period studied the subsidy per user and per month is estimated in 619€.

The managerial implications of the efficiency results are explored in detail for CYH 1, in 2013. The managerial efficiency score achieved for this CYH is 76.5% which is compared with the two benchmarks. For this inefficient CYH, it is also possible to obtain, as by-products of the DEA efficiency assessment, a set of targets for becoming efficient, as shown in Table 2. Targets for this institution indicate that it is possible to decrease the staff and operating costs by 23.5% without decreasing the social care services provided to the observed number of users. According to the policy of allocating subsidies, the regulator should reduce the observed subsidy by 31.2%, as shown in Table 2. Considering all the institutions, the proposed policy of allocating the subsidies enables a reduction by 24.5% of the total subsidies observed, which enables a saving of 2 173 565€ in the period analysed.

Table 2 Targets for CYH 1 in 2013

	Variable	Observed	Targets.
DEA model	Operating costs (€)	85 948	65 732
	Staff cost (€)	237 967	181 995
	Users	39	39
Proposed policy of subsidies	Subsidies (€)	420 900	289 692

5   Conclusions

The study purposes a DEA framework which can fulfil the interest of the SSI in managing and allocating public funds for CYH, which are more efficient in managing the subsidies received, in detriment of inefficient institutions. A DEA model is used to assess the managerial efficiency of NPO in minimizing the staff and operating costs required to provide the social services to the level of users in each institution. Based on the DEA results, a fair policy for allocating the subsidies is proposed according to real number of users and the cost levels observed on benchmarks. In the period studied, the subsidy per user and per month is estimated in 619€ being in the interval used by the regulator. Overall, the CYH should reduce their current operating and staff costs by 36%, on average. It is also observed that the institutions tend to decrease their managerial efficiency over time due to the increasing of the cost by user. This can be due to the increasing amount of the subsidies per user allocated by the regulator since 2012.

In future developments, the application of the proposed framework will be explored in data from CYH in other regions. The users satisfaction and their quality of life should also be investigated to complement the proposed approach.

References

[1] Taylor, M., Taylor, A.: Performance measurement in the Third Sector: the development of a stakeholder-focussed research agenda. Production Planning & Control 25(16), 1370–1385 (2013)

[2] Vakkuri, J.: Research techniques and their use in managing non–profit organisations – An illustration of DEA analysis in NPO environments. Financial Accountability & Management 19(3), 243–263 (2003)

[3] Sinuany-Stern, Z., Sherman, H.D.: Operations research in the public sector and non-profit organizations. Annals of Operations Research 221(1), 1–8 (2014)

[4] Liu, J.S., Lu, L.Y.Y., Lu, W.-M., Lin, B.J.Y.: A survey of DEA applications. Omega 41(5), 893–902 (2013)

[5] Berber, P., Brockett, P.L., Cooper, W.W., Golden, L.L., Parker, B.R.: Efficiency in fundraising and distributions to cause-related social profit enterprises. Socio-Economic Planning Sciences 45(1), 1–9 (2011)

[6] Carvalho, M.: Sistema nacional de acolhimento de crianças e jovens. In: Gulbenkian, F.C. (ed.) Programa Gulbenkian de Desenvolvimento Humano (2013)

[7] Sillanpää, V.: Performance measurement in welfare services: a survey of Finnish organisations. Measuring Business Excellence 15(4), 62–70 (2011)

[8] Polonsky, M., Grau, S.L.: Assessing the social impact of charitable organizations-four alternative approaches. International Journal of Nonprofit & Voluntary Sector Marketing 16(2), 195–211 (2011)

[9] Huang, H.J., Hooper, K.: New Zealand funding organisations. Qualitative Research in Accounting & Management 8(4), 425–449 (2011)

[10] Grigoroudis, E., Orfanoudaki, E., Zopounidis, C.: Strategic performance measurement in a healthcare organisation: A multiple criteria approach based on balanced scorecard. Omega 40(1), 104–119 (2012)

[11] Juaneda, E., González, L., Marcuelo, C.: El reto de la calidad para el Tercer Sector Social. Análisis de casos de implantación del modelo EFQM. Cuadernos de Gestión 13(2), 111–126 (2013)

[12] Medina-Borja, A., Triantis, K.: Modeling social services performance: a four-stage DEA approach to evaluate fundraising efficiency, capacity building, service quality, and effectiveness in the nonprofit sector. Annals of Operations Research 221(1), 285–307 (2014)

[13] Iparraguirre, J.L., Ma, R.: Efficiency in the provision of social care for older people. A three-stage data envelopment analysis using self-reported quality of life. Socio-Economic Planning Sciences 49, 33–46 (2015)

[14] Martin, L.L.: Comparing the performance of multiple human service providers using data envelopment analysis. Administration in Social Work 26, 45–60 (2002)

[15] Berber, P., Brockett, P.L., Cooper, W.W., Golden, L.L., Parker, B.R.: Efficiency in fundraising and distributions to cause-related social profit enterprises. Socio-Economic Planning Sciences 45(1), 1–9 (2011)

[16] ISS. Gestão da qualidade das respostas sociais: Lar de infância e juventude. Tipografia Peres, Lisboa (2007)

[17] Charnes, A., Cooper, W.W., Rhodes, E.: Measuring the efficiency of decision making units. European Journal of Operational Research 2(6), 429–444 (1978)

Distribution Supply Chain Inventory Planning under Uncertainty

Joaquim Jorge Vicente, Susana Relvas, and Ana Paula Barbosa-Póvoa

CEG-IST – Centre for Management Studies, Instituto Superior Técnico
Universidade de Lisboa, Portugal

Abstract. This paper considers a multi-echelon inventory/distribution system formed by N-warehouses and M-retailers that manages a set of diverse products in a dynamic environment. Transshipment at both regional warehouses and retailers levels is allowed. A mixed integer linear programming model is developed, where product demand at the retailers is assumed to be not known. The problem consists of determining the optimal reorder policy by defining a new concept of robust retailer order in a two level programming approach, which minimizes the overall system cost, including ordering, holding in stock and in transit, transportation, transshipping and lost sales costs. A case study based on a real retailer distribution chain is solved.

Keywords: Supply chain management, inventory planning, mixed integer linear programming, uncertainty, scenario planning approach.

1 Introduction

Multi-warehouse/multi-retailer/multi-product and multi-period distribution supply chains are complex systems through which flows of products have to be correctly managed along with the inventory policies practice within each entity. In order to overcome this challenging problem, the present paper addresses the inventory planning problem in such systems by using exact optimization methods.

Research on uncertainty can be classified by the type of approach used to represent it. Two types of approaches are identified: probabilistic approach and the scenario planning approach. The choice of the best method is context-dependent, with no single theory being sufficient to model all kinds of uncertainty [1]. In this paper, our focus is the scenario planning approach applied to a distribution supply chain (SC), where future demand is described through discrete scenarios.

Scenario planning approach attempts to capture uncertainty by representing it in terms of a moderate number of discrete realizations of the stochastic quantities, constituting distinct scenarios. The objective is to find robust solutions that perform well under all scenarios. Mohamed [2] used a scenario planning approach to decide on the design of a production and distribution network that operates under varying exchange rates. Tsiakis *et al.* [3] used a scenario planning approach to design a multi-echelon distribution supply chain network under demand uncertainty using a mixed-integer

© Springer International Publishing Switzerland 2015

A.P.F.D. Barbosa Póvoa and J.L. de Miranda (eds.), *Operations Research and Big Data*,
Studies in Big Data 15, DOI: 10.1007/978-3-319-24154-8_29

linear programming model. Combining scenario planning with supply-chain planning achieves the best of both worlds, which leads to long-term competitive advantage, as said in [4]. More recently, [5] and [6] adopted a scenario planning approach for handling the uncertainty in product demands. They proposed an optimal design of supply chain networks using a mixed-integer linear programming (MILP).

The scenario planning approach is a two level approach that has been extensively applied in design and planning problems, whereas we are focusing in an operational problem. Thus, a new concept is derived which is the robust ordering policy that should be adopted by each retailer. In this sense, the order quantity and timing has to be uncertainty free, but considering the different realizations of the uncertain parameters. The above identified opportunity is explored along the present paper where a generic model is proposed that determines both the inventory and distribution plans of the retailers and warehouses that minimize the distribution supply chain total costs under an uncertain demand environment.

2   Inventory Planning Mathematical Model under Uncertainty

The distribution supply chain inventory planning problem presented is formulated as a MILP model. The scenário approach used follows the work of [3], where demand can be described by a set of demand scenarios including both optimistic and pessimistic demands in a given time period.

We consider that retailers have own management autonomy and, therefore, have to plan their orders deterministically - robust retailer order. One order quantity placed by one retailer may be replenished through multiple flows.

The indices, constants, sets, parameters and variables used in the model formulation are defined using the following notation:

Indices
i - product; j, k, l, m - entity node; s - product demand scenario; t - time period

Constants
NP number of products; NW number of regional warehouses; NR number of retailers; NT number of time periods; NS number of product demand scenarios

Sets
$i \in P = \{1,2,...,NP\}$ products

$j,k,l,m \in I = \{0,1,2,...,NW,NW+1,NW+2,...,NW+NR\}$ SC nodes

$t \in T = \{1,2,...,NT\}$ time periods

$s \in S = \{1,2,...,NS\}$ scenarios for uncertain product demands

$W = \{1,2,...,NW\}, W \subset I$ regional warehouses

$R = \{1,2,...,NR\}, R \subset I$ retailers

$W_o = \{0\}, W_o \subset I$ central warehouse

$DN = \{1,2,...,NW, NW+1, NW+2,...,NW+NR\}, DN \subset I$ demand nodes

$SN = \{0,1,2,...,NW\}, SN \subset I$ supply nodes

Parameters

BGM a large positive number; CD_{ikst} customer demand of the product i at k in scenario s in time period t; HOC_{ij} unitary holding cost of the product i at j per time period; HTC_{ijk} unitary holding in transit cost of the product i from j to k; Ito_{ij} initial inventory level of the product i at j; LSC_{ijt} unitary lost sale cost of the product i at j in time period t; LTT_{jk} transportation lead time from j to k; OC_{ij} ordering cost of the product i at j; SS_{ij} safety stock level of the product i at j; STC_{jt} storage capacity at entity j in time period t; $TRACMAX_{jk}$ maximum transportation capacity from j to k; $TRACMIN_{jk}$ minimum transportation capacity from j to k; TRC_{ijk} unitary transportation cost of the product i from j to k ; φ_s probability of customer demand scenario s.

Non-Negative Continuous Variables

FI_{ijst} inventory of product i at j in scenario s at time period t; LS_{ijst} lost sales of product i at j in scenario s at time period t; RO_{ikt} robust retailer order of product i at k at time period t; SQ_{ijkst} shipping quantity of product i from j to k in scenario s in time period t.

Binary Variables

$BV1_{ijst} = 1$ if a regional warehouse order of product i is placed by j in scenario s in time period t, 0 otherwise; $BV2_{ikt} = 1$ if a robust retailer order of product i is placed by k in time period t, 0 otherwise.

The aim is to minimize the expected value of the total cost considering all scenarios and their probability of occurrence. This leads to the objective function (1).

$$Minimize\ total\ expected\ cost = \sum_{i \in P} \sum_{k \in R} \sum_{t \in T} OC_{ik} \times BV2_{ikt}$$

$$+ \sum_{s \in S} \varphi_s \times \left(\sum_{i \in P} \sum_{j \in W} \sum_{t \in T} OC_{ij} \times BV1_{ijst} + \sum_{i \in P} \sum_{j \in I} \sum_{t \in T} \left(HOC_{ij} \times FI_{ijst} + LSC_{ijt} \times LS_{ijst} \right) \right.$$

$$\left. + \sum_{i \in P} \sum_{j \in I} \sum_{k \in I} \sum_{t \in T} \left(\left(HTC_{ijk} \times LTT_{jk} + TRC_{ijk} \right) \times SQ_{ijkst} \right) \right) \tag{1}$$

The minimization of system costs is subject to the following constraints:

$$FI_{ijs1} = Ito_{ij} + SQ_{i,0,j,s,t-LTT_{0j} | LTT_{0j}=0} - \sum_{k \in R} SQ_{ijks1} - \sum_{l \in W \wedge l \neq j} SQ_{ijls1}$$

$$+ \sum_{l \in W \wedge l \neq j} SQ_{i,l,j,s,t-LTT_{lj} | LTT_{lj}=0}, i \in P, j \in W, s \in S, t=1 \tag{2}$$

$$FI_{ijst} = FI_{i,j,s,t-1} + SQ_{i,0,j,s,t-LTT_{0j}|LTT_{0j}<t} - \sum_{k \in R} SQ_{ijkst} - \sum_{l \in W \wedge l \neq j} SQ_{ijlst}$$

$$+ \sum_{l \in W \wedge l \neq j} SQ_{i,l,j,s,t-LTT_{lj}|LTT_{lj}<t}, i \in P, j \in W, s \in S, t \in T \setminus \{1\} \tag{3}$$

$$RO_{ikt} = \sum_{j \in W} SQ_{i,j,k,s,t-LTT_{jk}|LTT_{jk}<t}$$

$$+ \sum_{m \in R \wedge m \neq k} SQ_{i,m,k,s,t-LTT_{mk}|LTT_{mk}<t}, i \in P, k \in R, s \in S, t \in T \tag{4}$$

$$FI_{iks1} = Ito_{ik} + RO_{ik1} - (CD_{iks1} - LS_{iks1}) - \sum_{m \in R \wedge m \neq k} SQ_{ikms1}, i \in P, k \in R, s \in S, t = 1 \tag{5}$$

$$FI_{ikst} = FI_{i,k,s,t-1} + RO_{ikt} - (CD_{ikst} - LS_{ikst})$$

$$- \sum_{m \in R \wedge m \neq k} SQ_{ikmst}, i \in P, k \in R, s \in S, t \in T \setminus \{1\} \tag{6}$$

$$SQ_{i0jst} + \sum_{l \in W \wedge l \neq j} SQ_{iljst} \leq BGM \times BV1_{ijst}, i \in P, j \in W, s \in S, t \in T \tag{7}$$

$$\sum_{s \in S} \varphi_s \times \left(\sum_{j \in W} SQ_{ijkst} + \sum_{m \in R \wedge m \neq k} SQ_{imkst} \right) \leq BGM \times BV2_{ikt}, i \in P, k \in R, t \in T \tag{8}$$

$$\sum_{i \in P} FI_{ijst} \leq STC_{jt}, j \in DN, s \in S, t \in T \tag{9}$$

$$\sum_{i \in P} SQ_{ijkst} \leq TRACMAX_{jk}, j \in DN, k \in DN, j \neq k, s \in S, t \in T \tag{10}$$

$$TRACMIN_{jk} \leq \sum_{i \in P} SQ_{ijkst}, j \in DN, k \in DN, j \neq k, s \in S, t \in T \tag{11}$$

$$SS_{ij} \leq FI_{ijst}, i \in P, j \in DN, s \in S, t \in T \tag{12}$$

$$FI_{ijst}, LS_{ijst}, RO_{ikt}, SQ_{ijkst} \geq 0, i \in P, j \in I, k \in I, s \in S, t \in T \tag{13}$$

$$BV1_{ijst}, BV2_{ikt} \in \{0,1\}, i \in P, j \in W, k \in R, s \in S, t \in T \tag{14}$$

Objective function (1) involves: robust retailer ordering cost (first term) and regional warehouse ordering cost (second term); holding cost at both stages of the supply chain, regional warehouses and retailers and lost sales cost (third term); and, finally, the holding in transit cost and transportation cost (fourth term). The second, third and fourth terms are affected by scenarios s. For $t = 1$ the inventory of product i at regional warehouses j by product demand s is given by constraint (2). For the remaining time periods one should use constraint (3). A robust retailer order quantity is introduced though the definition of a new variable RO_{ikt} and another binary variable is needed for robust retailer order $BV2_{ikt}$. At retailers, the robust retailer order (shipment and transshipment) quantity is given by equation (4). For $t = 1$ the inventory of product i at the retailers k is given by constraint (5). Constraint (6) is applicable for the remaining time periods. If the transportation amount between two nodes is not zero, the binary variable $BV1_{ijst}$ equals 1, constraint (7). In constraint (8), the left hand side represents the robust retailer order quantity. The total inventory stored at any node must respect the storage capacity - constraint (9). At any time period t, the sum of the shipping quantity of each product i must respect the transportation minimum and maximum limits - constraints (10) and (11). Constraint (12) ensures that the inventory of each product i must be higher or equal than the required safety stock. The model uses non-negative continuous variables (13) and binary variables (14).

3   Inventory Planning Case Study

A Retail Company case study is here analyzed using the proposed inventory planning policy. The GAMS 23.5 modeling language combined with the CPLEX 12.2 were used to solve the problem in hand. An Intel CORE i5 CPU 2.27GHz and 4GB RAM was utilized. The stopping criteria were either a computational time limit of 7200 s or the determination of the optimal solution.

The distribution supply chain involves one central warehouse, two regional warehouses and four retailers. Three family types of products are considered. The maximum storage capacity of regional warehouses and retailers is of 500 units. The storage capacity of central warehouse is unlimited. Two transportation options are considered: transportation quantities between 0 and 500 units (SQ 0-500) and transportation quantities between 0 and 30 units (SQ 0-30). A seven period planning horizon is used. Lead time among retailers (for the transshipment operation) is assumed equal to one time period. The cost parameters are (in Euro) i) the ordering cost is 20, ii) the holding costs are 0.2 (warehouses) and 0.6 (retailers), iii) the holding in transit costs are 0.3 (from central warehouse to regional warehouses and transshipment between warehouses) and 0.9 (between warehouses and retailers or transshipment between retailers) and iv) the lost sales cost is 25. Tables from 1 to 4 present the remaining implemented parameters' values.

Table 1 Unitary product transportation costs (TRC) between entities (euro)

	W1	W2	R1	R2	R3	R4
W0	0.55	0.22	-	-	-	-
W1	-	0.7	0.22	0.2	0.32	0.38
W2	0.7	-	0.68	0.52	0.34	0.1
R1	-	-	-	0.2	0.8	1.3
R2	-	-	0.2	-	0.3	1.0
R3	-	-	0.8	0.3	-	0.36
R4	-	-	1.3	1.0	0.36	-

Three probabilities of product demand scenarios are considered, i. e., $\varphi_1 = 0.5$, $\varphi_2 = 0.3$, $\varphi_3 = 0.2$. The first product demand scenario is considered in table 3 – base scenario. The product demand for the second scenario, is a pessimist one, and for the third scenario, is an optimistic one, which present values lower and higher, respectively, than the base scenario. Due to space limitations, their values are not presented.

Table 2 Initial inventory level (Ito)/Safety Stock (SS) on warehouses and retailers (unit).

	W1	W2	R1	R2	R3	R4
Product1	45/14	30/14	24/7	22/3	20/3	18/3
Product2	15/11	11/11	16/2	14/2	12/2	10/2
Product3	11/8	9/8	8/2	4/2	6/1	9/1

Table 3 Customer Demand at the retailers of product1/product2/product3 in each period (unit).

	R1	R2	R3	R4	Total
Period1	12/4/3	11/3/7	14/6/9	3/1/2	40/14/21
Period2	10/12/2	9/11/9	6/1/0	15/9/5	40/33/16
Period3	14/2/4	13/1/1	9/11/1	9/5/1	45/19/7
Period4	8/14/1	7/13/3	11/2/6	8/4/6	34/33/16
Period5	16/8/7	15/9/4	20/10/2	10/6/8	61/33/21
Period6	6/10/5	5/5/5	0/5/3	7/2/9	18/22/22
Period7	18/6/6	17/7/2	10/7/6	11/8/3	56/28/17
Total	84/56/28	77/49/31	70/42/27	63/35/34	294/182/120

Table 4 Lead Time Transportation (LTT) of product (time period).

	W1	W2	R1	R2	R3	R4
W0	2	1	-	-	-	-
W1	-	1	1	1	1	1
W2	1	-	2	2	1	0

The aim of such study is to propose an inventory planning that can handle with all three scenarios while minimizing the total expected cost. The most representative results are shown in table 5, where the total expected cost and the global service level are illustrated. It can be seen that the total expected cost grows from SQ 0-500 to SQ 0-30. For the global service level, generally SQ 0-500 returns a higher service level than SQ 0-30.

Total lost sales per product demand scenario and per product, for both settings for case study under uncertain product demands are presented in table 6. The optimistic product demand scenario (s3) leads to the worst results, as it has the higher number of lost sales. The pessimistic product demand scenario (s2) results lead to be best lost sale solution.

Table 5 Total expected cost and global service level per retailer and per product1/ product2/ product3 (aggregated on time horizon) under uncertain product demands.

	SQ 0-500	SQ 0-30
Total expected cost	2398.06	2598.90
Retailer 1	1/0.99/0.98	1/0.98/0.97
Retailer 2	1/0.99/0.82	0.98/0.99/0.82
Retailer 3	1/0.99/0.85	1/0.99/0.82
Retailer 4	1/1/1	1/1/1

Table 7 shows the total robust retailer order (shipment and transshipment) quantity per product demand scenario and per product, which is aggregated on time horizon and on retailers' echelon, for both transportation options under uncertain product demands. The total robust retailer order quantity is scenario independent. However, the shipment quantity (from the regional warehouses) and the transshipment quantity (from the others retailers) are scenario dependent. Note that both transportation quantities (shipment and transshipment) are capacity limited. In general, the pessimistic scenario (s2) makes more use of transshipment than the others. Regarding the total number of robust retailer orders per product1/product2/product3 these are 11/7/7 and 14/11/9 respectively for SQ 0-500 and SQ 0-30. It is an expected result, since with more constrained distribution flows, orders sizes are more restricted leading to the need of being replenished through a higher number of orders.

Table 6 Total lost sales per product demand scenario and per product1/ product2/ product3 (aggregated on time horizon and on retailers' echelon) (unit).

	SQ 0-500	SQ 0-30
s1	0/0/9	1/0/10
s2	0/0/5	0/0/5
s3	0/7/24	5/11/28

Table 7 Total robust retailer order (shipment and transshipment) quantity per product demand scenario and per product1/ product2/ product3 (aggregated on time horizon and on retailers' echelon) under uncertain product demands (unit).

	SQ 0-500			SQ 0-30		
	shipment	transshipment	total	shipment	transshipment	total
s1	226/138/90	78/74/58	304/212/148	225/138/91	56/50/47	281/188/138
s2	172/85/54	132/127/94	304/212/148	172/93/58	109/95/80	281/188/138
s3	282/187/131	22/25/17	304/212/148	277/183/127	4/5/11	281/188/138

4   Conclusions

This paper proposes a generic inventory planning model for a multi-period/multi-warehouse/multi-retailer/multi-product distribution supply chain. An inventory and distribution plan is obtained, which minimizes the total costs under demand uncertainty through the determination of a robust retailer order. The definition of the robust retailer order concept can be considered the main contribution of this paper.

Future research should include further detailed validation of the proposed model through the study of more complex distribution supply chain structures over larger periods of time with more products. The definition of the safety stock levels and the comparisons between centralized and decentralized systems under uncertainty accounting for risk pooling advantages should also be the focus of further research.

References

[1] Zimmermann, H.J.: An application-oriented view of modeling uncertainty. European Journal of Operational Research 122, 190–198 (2000)
[2] Mohamed, Z.M.: An integrated production-distribution model for a multi-national company operating under varying exchange rates. International Journal of Production Economics 58, 81–92 (1999)
[3] Tsiakis, P., Shah, N., Pantelides, C.C.: Design of multi-echelon supply chain networks under demand uncertainty. American Chemical Society 40, 3585–3604 (2001)
[4] Sodhi, M.S.: How to do strategic supply-chain planning. MIT Sloan Management Review 45, 69–75 (2003)
[5] Georgiadis, M.C., Tsiakis, P., Longinidis, P., Sofioglou, M.K.: Optimal design of supply chain networks under uncertain transient demand variations. Omega 39, 254–272 (2011)
[6] Cardoso, S.R., Barbosa-Póvoa, A.P., Relvas, S.: Design and planning of supply chains with integration of reverse logistics activities under demand uncertainty. European Journal of Operational Research 226, 436–451 (2013)

Author Index

Printed in the United States
By Bookmasters